3D Printing Blueprints

Design successful models for home 3D printing, using a Makerbot or other 3D printers

Joe Larson

BIRMINGHAM - MUMBAI

3D Printing Blueprints

First published: August 2013

Production Reference: 1160813

Published by Packt Publishing Ltd.
Livery Place
35 Livery Street
Birmingham B3 2PB, UK.

ISBN 978-1-84969-708-8

www.packtpub.com

Cover Image by Joseph Larson (joealarson@gmail.com)

Credits

Author
Joe Larson

Reviewers
Henry Garner
Andrew Mazzotta
Thomas P.McDunn
Erwin Ried

Acquisition Editor
Edward Gordon

Lead Technical Editor
Arun Nadar

Technical Editors
Shashank Desai
Dennis John
Chandni Maishery
Sanhita Sawant

Project Coordinator
Shiksha Chaturvedi

Proofreader
Mario Cecere

Indexer
Monica Ajmera Mehta

Production Coordinator
Shantanu Zagade

Cover Work
Shantanu Zagade

About the Author

Joe Larson is one part artist, one part mathematician, one part teacher, and one part technologist. It all started in his youth on a Commodore 64 doing BASIC programming and low resolution digital art. As technology progressed, so did Joe's dabbling eventually taking him to 3D modeling while in high school and college, momentarily pursuing a degree in Computer Animation. He abandoned the track for the much more sensible goal of becoming a Math teacher, which he accomplished when he taught 7th grade Math in Colorado. He now works as an application programmer.

When Joe first heard about 3D printing, it took root in his mind and he went back to dust off his 3D modeling skills. In 2012, he won a Makerbot Replicator 3D printer in the Tinkercad/Makerbot Chess challenge with a chess set that assembles into a robot. Since then his designs on Thingiverse, have been featured on Thingiverse, Gizmodo, Shapeways, Makezine, and others. He currently maintains the blog `joesmakerbot.blogspot.com`, documenting his adventures in 3D printing.

Dedicated to my wife, who I've seen far too little of during the process of making this book.

Thanks to the people at Packt Publishing who reached out to help me write this book.

Special thanks to the awesome people at Makerbot.

About the Reviewers

Henry Garner started 3D printing in 2010 after buying a MakerBot Cupcake CNC kit as a present for himself. Then a professional software developer with no 3D modeling skills, the obvious next step was to learn the printer's wire protocol and control the print head directly with his own code. The result was the open source Ruby library called Cupcake and many tangled knots of extruded plastic.

It was whilst studying for a Fine Art degree years earlier that he first became interested in programming as a means to create interactive installations and reactive sculptures. The combination of technology and tangible objects remains his passion, and he thinks 3D printing offers a fantastic new way to bring ideas out of the confines of a computer screen into the physical world.

Henry is now Chief Technologist at Likely, a big data analytics company based in the heart of East London's Tech City. When not working, he is often to be found at his art studio by the Tate Modern, floor still littered with extruded plastic tumbleweed.

You can follow him on Twitter at @henrygarner.

Andrew Mazzotta started his career in finance with an MBA. After recently traveling the world, 70 countries in three years, he changed his career for engineering and is now working on three degrees in mechanical engineering, electrical engineering, and computer science. He is currently (August 2013) building RepRaps in Albania for undeveloped areas. The project is in collaboration with Printers for Peace.

He started www.3dhacker.com, a free site dedicated to 3D printing. Members can showcase 3D printers, extruders, printer upgrades, STL model designs, software, printing tutorials, and so on. Additionally, there is a forum for members to support their products/services and a blogging platform for all members to use as well.

> I would like to thank all the people that have made 3D printing available to the less fortunate!

Thomas P. McDunn is an engineer and tinkerer and finds 3D printing fuels an ever growing list of projects and experiments. Though formally trained in Mechanical Engineering, receiving a Bachelor's and a Master's Degree from University of Wisconsin-Madison, his interest in computers pulled him on the fence between Mechanical and Electrical Engineering. He constructed his own microcomputer for home experimenting and cut his teeth on electromechanical applications of microprocessors at a time when memory was expensive and code had to be small, not only for memory considerations but for execution speed and hand coding sake. Applying knowledge of mechanical and electrical systems with servo control theory he developed a career in motion control in the Machine Tool Industry and was granted a patent in 1992 for a microprocessor-based transfer line control. He experimented with robotics with a Hero 2000 robot and worked briefly in the industrial robotics industry. Thomas enjoys education and has developed many hands-on curricula for quick immersion of technical concepts. More recently, Thomas has applied his managerial and marketing experience and consults with small businesses in leveraging social media as a marketing tool. Frustrated with the hodge-podge of image sizes used in social media he self-published a book, "72 Pixels" that details the image size requirements of the most popular social media applications. Spurred by an episode of "The Shark Tank", he started a blog to pass along lessons learned in inventor and investor relations. Thomas spends a lot of time online and is enamored with the growth and accomplishments of open source projects.

The open source movement has paved the way for many innovations and more to come. Thomas studies open source applications in a wide arena including Arduino, GIMP, Inkscape, REPRAP, and Drones to name a few. Thomas has experience of many types of 3D printing and rapid prototyping models, setting up a design, and prototyping bureau including SLA, SLS, FDM, Zprint, and PolyJet machines. He has a Makerbot and has recently added a Rostock Max to his personal 3D printer arsenal. Thomas is constantly on the lookout for world changing applications of 3D printing technology. He recently registered with Robohand, a website and organization that makes affordable prosthetics available for children who are born without fingers using 3D printing technology. He catalogs his personal printing projects at `www.tpmtech.biz/Makerbot`.

I'd like to thank my wife Holly for her encouragement and patience in supporting my technological passions.

Erwin Ried has been enjoying computers and electronics since the age of seven, when he first received his Atari 800 XE. Playing with the buggy coding examples from the Atari booklet always sparked something special in his mind; the idea of governing the machine.

In 2009, HP and Microsoft chose his website as one of the top 50 world best blogs in the HP Magic Giveaway. Later in 2011, LG electronics selected his invention SinStandby as the best green energy related solution for "Casa Eficiente del Siglo XXI".

Now, he is a Computer Civil Engineer (Bachelor) from Chile who loves any kind of challenge including electronics, mechanics, and coding in particular when they involve 3D Printing, electronics, robotics, automation, games and/or programming.

www.PacktPub.com

Support files, eBooks, discount offers and more

You might want to visit www.PacktPub.com for support files and downloads related to your book.

Did you know that Packt offers eBook versions of every book published, with PDF and ePub files available? You can upgrade to the eBook version at www.PacktPub.com and as a print book customer, you are entitled to a discount on the eBook copy. Get in touch with us at service@packtpub.com for more details.

At www.PacktPub.com, you can also read a collection of free technical articles, sign up for a range of free newsletters and receive exclusive discounts and offers on Packt books and eBooks.

http://PacktLib.PacktPub.com

Do you need instant solutions to your IT questions? PacktLib is Packt's online digital book library. Here, you can access, read and search across Packt's entire library of books.

Why Subscribe?

- Fully searchable across every book published by Packt
- Copy and paste, print and bookmark content
- On demand and accessible via web browser

Free Access for Packt account holders

If you have an account with Packt at www.PacktPub.com, you can use this to access PacktLib today and view nine entirely free books. Simply use your login credentials for immediate access.

Table of Contents

Preface

A new industrial age is here. Machines designed to build useful and interesting objects have moved from the factory to the home. But these 3D printers can't make things without a design. Whether you have a 3D printer or not, designing things for 3D printers to make is the best way to become a part of the 3D printing movement. Learn to design successful models for home by 3D printing on a Makerbot or other 3D printer with cool hands-on lessons.

If you've ever won a round of Pictionary you've got all the artistic skill it takes to get started. If you've ever gotten past level 1 on Tetris then you've got spatial reasoning. If you've ever played with modeling clay then you know all about designing in three dimensions. You can learn and practice the rules of design that will take your virtual models to real life prints you can hold in your hands as well as enable your creations to stand out on popular websites such as Thingiverse.

This book uses blueprints; simple, fun projects that teach Blender modeling for 3D printing in hands-on lessons. First you'll learn basic modeling and make a small simple object. Then each new project brings with it new tools and techniques as well as teaching the rules of 3D printing design. Eventually you'll be building objects designed to repair or replace everyday objects. Finally, you'll be able to even tackle other people's models and fix them to be 3D printable.

What this book covers

Chapter 1, Design Tools and Basics, will start with the rules of designing objects for successful 3D prints and then introduce the software that will be used.

Chapter 2, Mini Mug, introduces the most common modeling tools to make a simple object.

Chapter 3, Face Illusion Vase, uses a reference image, a picture, to help create the shape of a 3D object.

Chapter 4, SD Card Holder Ring, takes measurement of real-life objects and translates them to the design space. Success is measured by how closely the print matches the real life object.

Chapter 5, Modular Robot Toy, combines separate parts with joints to make a single object.

Chapter 6, D6 Spinner, uses the add-on functionality to allow Blender to create new objects and using that to model a new way to choose a number from 1 to 6.

Chapter 7, Teddy Bear Figurine, introduces a different method of modeling that can be used to make appealing organic shapes.

Chapter 8, Repairing Bad Models, is a good skill to have when working with other's 3D models, particularly those that might not have been made for 3D printing.

Chapter 9, Stretchy Bracelet, shows how advanced 3D printing options can change the way a model is printed.

Chapter 10, Measuring – Tips and Tricks, are important to know when translating real life into the design space.

What you need for this book

This book uses only Blender for 3D modeling available at www.blender.org, a free, open source program. The first chapter covers downloading and installing Blender.

If you have a Makerbot or other brand of 3D printer you will need software to prepare models for print. Either Slicer or ReplicatorG is recommended for Makerbots. No other software will be necessary for this book.

An account on Thingiverse (www.thingiverse.com) to upload your own models is recommended but not required.

Who this book is for

This book is for anyone with an interest in 3D printing and the slightest bit of computer skill. Whether you own a 3D printer or not you can design for them. All it takes is some free software, this book, and a little creativity and someday you'll be able to hold something designed on a computer in your hands. No special computer skills are necessary beyond simple file and directory navigation. No previous 3D modeling experience is necessary at all.

Conventions

In this book, you will find a number of styles of text that distinguish between different kinds of information. Here are some examples of these styles, and an explanation of their meaning.

Code words in text, database table names, folder names, filenames, file extensions, pathnames, dummy URLs, user input, and Twitter handles are shown as follows: "Type in `MakerbotBlueprints` as the name for the new directory."

New terms and **important words** are shown in bold. Words that you see on the screen, in menus or dialog boxes for example, appear in the text like this: "Click on the **Download** link".

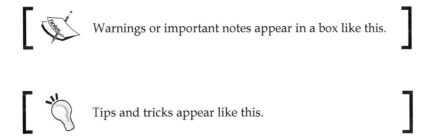

> Warnings or important notes appear in a box like this.

> Tips and tricks appear like this.

Reader feedback

Feedback from our readers is always welcome. Let us know what you think about this book—what you liked or may have disliked. Reader feedback is important for us to develop titles that you really get the most out of.

To send us general feedback, simply send an e-mail to `feedback@packtpub.com`, and mention the book title via the subject of your message.

If there is a topic that you have expertise in and you are interested in either writing or contributing to a book, see our author guide on `www.packtpub.com/authors`.

Customer support

Now that you are the proud owner of a Packt book, we have a number of things to help you to get the most from your purchase.

Downloading the color images of this book

We also provide you a PDF file that has color images of the screenshots/diagrams used in this book. The color images will help you better understand the changes in the output. You can download this file from: `http://www.packtpub.com/sites/default/files/downloads/7088OT_ColoredImages.pdf`.

Errata

Although we have taken every care to ensure the accuracy of our content, mistakes do happen. If you find a mistake in one of our books—maybe a mistake in the text or the code—we would be grateful if you would report this to us. By doing so, you can save other readers from frustration and help us improve subsequent versions of this book. If you find any errata, please report them by visiting `http://www.packtpub.com/submit-errata`, selecting your book, clicking on the **errata submission form** link, and entering the details of your errata. Once your errata are verified, your submission will be accepted and the errata will be uploaded on our website, or added to any list of existing errata, under the Errata section of that title. Any existing errata can be viewed by selecting your title from `http://www.packtpub.com/support`.

Piracy

Piracy of copyright material on the Internet is an ongoing problem across all media. At Packt, we take the protection of our copyright and licenses very seriously. If you come across any illegal copies of our works, in any form, on the Internet, please provide us with the location address or website name immediately so that we can pursue a remedy.

Please contact us at `copyright@packtpub.com` with a link to the suspected pirated material.

We appreciate your help in protecting our authors, and our ability to bring you valuable content.

Questions

You can contact us at `questions@packtpub.com` if you are having a problem with any aspect of the book, and we will do our best to address it.

1
Design Tools and Basics

Owning a Makerbot 3D printer means being able to make anything you want at a push of a button, right? 3D printer owners quickly find that while 3D printers have no end of things they can produce, they also are not without their limitations. Designing an object without 3D printing in mind will result in a failed print that more resembles a bird nest or a bowl of spaghetti.

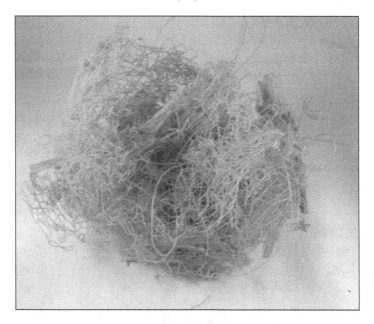

Making a 3D printable object requires learning a few rules, some careful planning, and design. But once you know the rules the results can be astounding. 3D printers can even produce things with ease that traditional manufacturing cannot, for example, objects with complex internal geometry that machining cannot touch.

There are many places online such as Makerbot's own Thingiverse that hosts a daily growing library of printable objects. Printing out other people's designs is all well and good for a while, but the most exciting part about 3D printing is that it can produce your designs and models. Eventually, learning how to model for 3D printing is a must.

Can you learn 3D modeling? If you've ever won a round of Pictionary you've got all the artistic skill it takes to get started. If you've ever gotten past level 1 on Tetris then you've got spatial reasoning. If you've ever played with modeling clay then you know all about designing in three dimensions.

Design basics

There are some design rules and basic ideas that will be true regardless of the modeling software used.

The working of 3D printing

3D printing has come a long way in terms of technology and cost allowing home 3D printers to be a reality. In this process there have been choices that will limit what can be printed. Seeing a 3D printer in action is the best way to learn about the process. Fortunately there are many 3D printing time lapse videos online of printers in action that can be found with a simple search.

3D printers build an object layer-by-layer from the bottom to the top. Plastic filament is heated and extruded, and each layer is built upon the last one. Usually the outside of the object is drawn and sometimes additional shells are added for strength. Then the inside is usually filled with a lattice to save plastic and provide some support for higher layers, however the inside is mostly air. This continues until the object is complete as shown in the following screenshot:

Because of this layer-by-layer process, if a design is made so that any part has nothing underneath it, dangling in the air, then the printer will still extrude some plastic to try to print the part which will just dangle from the nozzle and be dragged into the next area where it will build up an ugly mess and ruin the print:

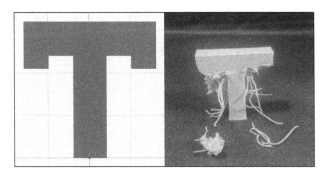

Building for supportless prints

One way of fixing the dangling object problem is to configure the preparing software to build the model "with supports". This means the slicer will automatically build a support lattice of plastic, up to the dangling part so that it has something to print on. Higher-end printers can actually print with a different material that can be dissolved away, but so far most home printers only use break-away supports. Either way after the print is complete it is left to the user to clean up this support material to extract the desired part.

While supports do allow the creation of objects that would be impossible any other way, the supports themselves are a waste of material and often don't remove cleanly leading to a messy bottom surface where they contact the print. If a part is designed needing supports that are hard to remove, such as if they're internal and partially obscured, it can be difficult and frustrating to completely remove the support material (this can be true for even the higher-end 3D printers). The process of removing it may actually damage the print.

It is possible and very easy with just the slightest application of cleverness to make designs that are printable without the need for any supports. So the blueprints in this book focus on making designs that print without supports. The limitations imposed by this demands just a little more effort but allow for the teaching of principles that are generally good to know.

Designing for dual extruders

Some models of Makerbot and other 3D printers have the ability to print in multiple colors at once using two different extruder heads feeding plastic from two different spools. There are some fun prints that come from this process. But as most Makerbots and other brands of home 3D printers do not have dual extruders at this time this book will not explore this process in detail. The basic idea of the process is creating two files that are aligned to print in the same space and combining them in the slicer.

Designing supportless – overhangs and bridges

When designing for supportless printing the rules are simple: Y prints, H prints okay, T does not print well.

Branching out with overhangs

It is possible to have the current layer slightly larger than the previous layer provided the overhang is not more than 45 degrees. This is because the current layer will have enough of the previous layer to stick to. *Hence a shape like the capital letter Y will successfully print standing up.*

However, if the overhang is too great or too abrupt the new layer will droop causing a print fail, hence a shape like the capital letter T does not print. (If the T is serif and thus has downward dangling bits, it will fail even worse, as illustrated previously.) So it is important to try to keep overhangs within a 45 degree cone as they go upwards.

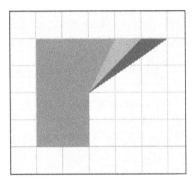

Building bridges

If a part of the print has nothing above it, but has something on either side that it can attach to, then it may be able to bridge the gap. But use caution. The printer makes no special effort in making bridges; they are drawn like any other layer: outline first, then infill. As long as the outline has something to attach to on both sides it should be fine. But if that outline is too complex or contains parts that will print in mid-air, it may not succeed. Being aware of bridges in the design and keeping them simple is the key to successful bridging. Even with a simple bridge some 3D printers need a little bit more calibration to print it well. Hence a shape like the capital letter H will successfully print most of the time.

Of course this discussion is purely illustrative of the way overhangs work or fail. In real life if a Y, H, or T needed to be printed the best way to do it would be to lay them down. But for purposes of illustration it still stands that Y prints, H prints okay, T does not.

Choosing a modeling tool

There are many choices of modeling programs that can be used to produce 3D printable objects. There are many factors including versatility, simplicity, and cost to take into account. A tool with too steep a learning curve can turn off new users to the idea all together. A tool with too limited a set of tools can frustrate a user when they hit the limit. Investing a lot of money into something that doesn't end up going anywhere can be extremely disappointing. So it is important to explore the options.

SolidWorks (`www.solidworks.com`) and other drafting oriented programs can do technical shapes with extreme precision. They include the necessary tools to accurately describe a shape that can be brought into the real work with high fidelity. However these sorts of tools tend to be costly and don't do artistic or organic shapes very well. Their highly technical nature also gives them a steep learning curve.

OpenSCAD (`www.openscad.org`) is free and famous among the people who make 3D printers and can make technically accurate models as well. OpenSCAD also allows the models to be parametric, meaning that by changing a few variables and recalculating a new shape is generated. But OpenSCAD is difficult to use unless the user has a very technical and programmatic mind since the shapes are literally built from lines of code.

Zbrush (`pixologic.com/zbrush`), Sculptris (`pixologic.com/sculptris`), or Wings3D (`www.wings3d.com`) are great tools for modeling organic shapes like the kind used in video games or animation. Sculptris and Wings3D are free and are very easy to pick up and use. But these tools lack when precision is necessary.

Sketchup (`www.sketchup.com`) is a great free program with a library of shapes built in ready to import and play with. Its modeling tools are great for precise or architectural models. Sketchup doesn't do organic shapes well either and it can be tricky loading the plug-ins necessary for Sketchup to export their models to something printable. Even then models from Sketchup often have to go through an extensive clean up phase before they'll be ready to print.

Autodesk 123D (`www.123dapp.com`) is not one but a whole suite of free programs designed around 3D modeling with specific focus on 3D printing. There are programs to design creatures or precise shapes. There is even an app for converting pictures of real life objects into 3D models. Some are programs that run in browser, some are downloads and some are apps for Apple devices. It's an eclectic and powerful group of programs. The Autodesk 123D suite's weakness is in its general immaturity. Autodesk is making great efforts to make modeling for 3D printing accessible for everyone but its tools still need to mature somewhat before they'll be ready to explore in depth.

Blender (`blender.org`) is a 3D animation program that features a robust set of modeling tools. Good for artistic and organic shapes, it can also be used when precision is needed. On top all that it is free and open source, so it is still in constant development. If Blender doesn't have a particular feature it is only a matter of time until it will be added. If fact by the time this book is published chances are the version of Blender used to make it will already be out of date, but most of Blender's functionality remains unchanged version-to-version. Blender is also completely customizable so that every feature, from the key strokes used to the overall look, can be changed. The downside of Blender is that its user interface is somewhat unintuitive. This causes Blender to have a famously difficult learning curve

Because of Blender's versatility and availability it is the tool of choice for the beginning 3D designer and this work.

Installing Blender

This will be the first project in the book. Fortunately downloading and installing Blender is as easy as 1-2-3-4.

1. Go to `www.blender.org`.
2. Click on the **Download** link.

3. Choose and download the installer for your system: Windows, Mac, or Linux, 32 bit or 64 bit.
4. Run the installer.

The installer will guide the process of loading Blender and adding icons to the system.

Windows Blender.org offers installer executable and ZIP files. The zip files are for advanced users who want a portable version of Blender. When in doubt choose the executable since it will set up icons making for easy access. If in doubt whether to use the 32 or 64-bit versions picking the 32-bit will insure compatibility, but it is a good idea to find out what type of system it is being installed on as 64-bit offers significant performance improvements.

Windows 7 or greater will confirm that the installer should be run. Click on **Yes** to assure Windows that it's okay to install Blender.

Then the installer will run. The install wizard's defaults are fine for most users. Simply put the mouse over the **Next** button and click on every button that appears under it. On the second screen read over the Blender Terms of Service and click on **I Agree** to proceed. Unless you manage your installed programs directories yourself it is best to leave the defaults on the third screen as it is. Then click on **Next** and the install process will start.

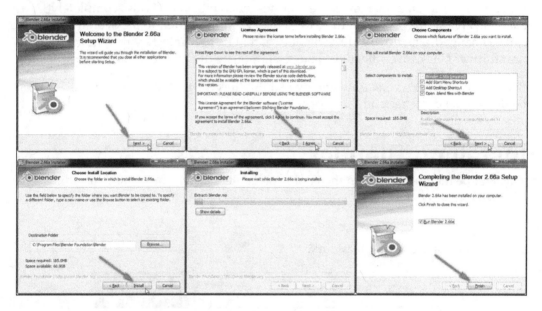

When the install process finishes leave the check box check and click on **Finish** to exit the installer and run Blender.

Getting acquainted with Blender

When Blender starts up, the Splash Screen can be dismissed by left-clicking outside the screen.

 The screenshots in this book use a custom color palette for print and a smaller window to minimize wasted screen space. Customizing the color palette will be discussed briefly later but these cosmetic changes will not affect the instructions presented at all.

The Blender interface is broken into several different customizable panels to keep things organized.

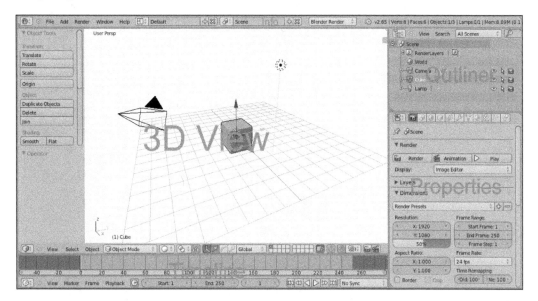

Each panel has resizing widgets in the upper-right and lower-left corners. By clicking on these widgets the panels can be split to add more panels or expanded into the territory of another to collapse panels at the user's preference. However, for now the default panels will be discussed as they provide the most common functionality for beginners.

The 3D View panel

The main window where things will be happening is the 3D View panel. The largest portion of the 3D view consists of the viewport where most of the work will take place.

On the left-hand side of the 3D View panel is a tool bar that consists of tools relevant to the selection and mode. If the tool bar is ever not visible it can be revealed (or hidden again) by pressing *T* or by clicking on the plus icon on the right-hand side that will appear when the toolbar is hidden. The specific tools in this bar will be explored as they are needed in the projects.

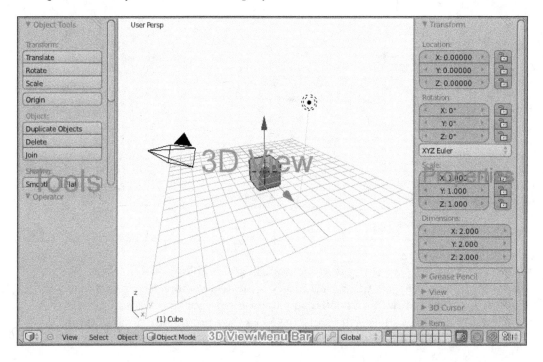

There is another plus icon on the right that will bring up the viewport properties with properties relevant to the current selection or the viewport. This can also be revealed or hidden by pressing the *N* key. Again, the specifics will be explored further as needed.

At the bottom of the 3D View panel is the 3D view menu bar with additional options followed by menus and icons related to editing and views. Hovering the mouse over each button will show what they are for.

The Outliner panel

The Outliner panel contains a hieratical view of all the objects in the scene. Each object can be selected by clicking on their name or the object can be hidden, locked, or excluded from rendering.

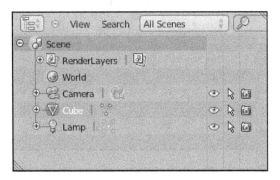

 Rendering means making a high quality picture from a scene for things such as animation or presentations. Doing a proper render includes setting up scene lights, cameras, textures, material properties, and many other functions that will not be explored in this book as it does not do anything that helps produce models for printing. However, exploring this functionality outside of this book can be good when trying to show off the models if printing them is not an option.

The Properties panel

The Properties panel is broken up into many tabs indicated by small icons. Hovering over the icons will show the name of the tab. For modeling the two tabs that will be used the most are the object and modifier tabs. Specific exploration of the tools contained therein will be done as needed.

The Info panel

On the top of the Blender windows is the Info panel. On the left-hand side of the Info panel there is an easy to navigate menu similar to the menu in most applications. This menu can be collapsed by clicking on the + button next to it and expanded by clicking the same. On the far right of the Info panel there is data about the current scene. If the data cannot be seen, hover the mouse over the panel, and use the middle scroll wheel or click-and-hold the middle mouse button (pressing the wheel like a button) and moving the mouse to pan the panel until the desired data is visible.

The Timeline panel

The Timeline panel is only relevant to doing animation and can be effectively ignored or collapsed for the purposes of this book.

 Because this book is only using a limited subset of Blender's functionality some things such as the Timeline panel could be customized away. However, since it is not the focus of this book to tell the reader how to customize their version of Blender, and because Blender has a much broader application, the screenshots that follow will have the Timeline panel visible. The reader is encouraged to explore Blender's other functionalities such as rendering and animation at their leisure and desire.

Proper stance

While all of Blender's functions are available from buttons and menus on the screen, typical Blender users rely heavily on hotkeys and shortcuts. Already the *T* and *N* keys have been discussed to bring up or collapse the **Tools** and **Properties** tabs in the 3D view. For this reason it is recommended that to use Blender have one hand on the mouse and the other hand on the keyboard at all times. This tends to be a common stance for many people but is mentioned for the few for whom learning this will be of great help.

Blender customization

One of Blender's strengths is its customizability. Almost every feature from the look and color, right down to the keystrokes and hotkeys are used for every action in Blender. Customization is accomplished in the **Properties** menu accessed from the **File | User Preferences** menu.

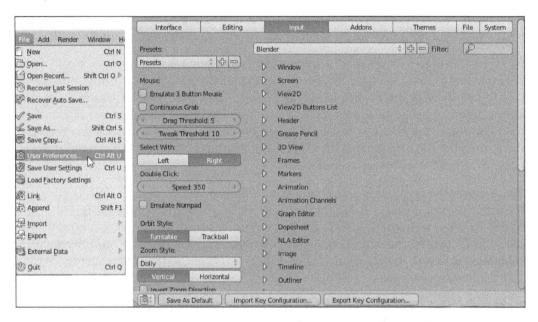

The buttons across the top, switch to the various categories. Each category is packed with options. A full exploration of these options is beyond the scope of this book but the reader is encouraged to explore these options and make Blender their own.

For instance if the reader is using a setup where a middle mouse button is unavailable, Blender contains an option to emulate the middle mouse button by pressing *Alt* along with *left-click*. Other systems may require other accommodations many of which are available in this menu.

Setting up for Mac OSX

Mac OSX users require special consideration. Blender is made for a three button mouse. If a single button mouse is all that is available, click when the instructions say **Left-Mouse Button**, use *Alt* with click for **Middle-Mouse Button**, and press *command* with click for **Right-Mouse Button**.

General Blender tips

Blender employs some conventions that are unique to its environment and as such getting acquainted with its most common quirks early can avoid frustration.

First and perhaps most importantly, *Ctrl + Z* for undo works in Blender will undo a multitude of mistakes. Undo in Blender remembers many past steps allowing backing up to a point before a grievous error was made. Remembering this when following along with the blueprints that follow will save the reader much frustration.

Next, the location of the mouse pointer is important when using hotkeys. For example the *T* and *N* keys for the **Tools** and **Properties** tabs do not bring up those tabs if the mouse pointer is not hovering over the 3D View. If the mouse is hovering over a different panel the reaction could be unpredictable. Pressing the *A* key with the mouse over the 3D View will toggle selection of all objects in the scene. Pressing the *A* key with the mouse over the **Object Tools** tab will collapse the expandable menu hiding all the options of that menu.

Blender uses the right-click on the mouse to select objects by default. This is perhaps the most counter intuitive thing for first time users, particularly because it will be encountered so frequently. But not everything has been swapped, just object selection. This behavior can of course be customized. If the reader would like to customize selection to the left mouse button then it is left to them to adjust the instructions accordingly.

Finally, the relation of Blender units to real life units is not by default defined in Blender. Generally it is just easier to remember that 1 grid point will translate to 1 millimeter in the printed object. As the scale is increased Blender inserts darker grid lines every 10 grid lines by default which correlate to centimeters. So the default cube in the default scene would measure 2 mm on each side, which is less than 1/10 of an inch, which is very small.

Suggested shortcuts

In the projects in this book, when a new idea is introduced it will first be introduced with detailed steps. Once a process is taught the next time the name of the operation and the shortcut for that process will be all that is given. For instance the first time the scale operation is introduced in *Chapter 2, Mini-mug*, the process is laid out including how to start, modify, and end the operation, but later an instruction like "**scale** (*S*) the object" is all that will be given. This does not mean that the keyboard shortcuts are the only way to do an operation but they are often the preferred method for experienced Blender users. The reader is free to accomplish the operation in any way that is comfortable for them.

The blueprints

This book has been designed to teach 3D printing design in a hands-on approach. A series of projects or blueprints will be presented and each one will introduce new tools and techniques. Each one builds on the last. Despite being a "virtual" process, 3D modeling has a surprisingly muscle memory aspect to it. The movements and processes need to be more than a mechanical process being executed, they need to be practiced so they can be fluid and eventually seamless. To that end the reader is advised not to skip any of the blueprints and follow along actually doing each one.

The objects being designed in this book are, most of them, very small so that they can be printed without taking too much time or producing too much waste. The reader can make larger versions if they like but that is left for their own challenge activities.

Summary

3D printing is cool. Learning to design your own models is the best way to take full advantage of 3D printing today. This book will teach 3D modeling by a series of hands-on activities so it's a good idea not to skip and actually follow along with each blueprint.

While home 3D printers have the capability to do break away supports these are messy and wasteful. It is possible to design things to be able to print without the need of any supports. When designing things for support-less 3D prints remember Y prints, H prints okay, T does not print. Keep outward inclines gradual and no more than 45 degrees to be safe.

There are many 3D modeling programs to choose from. Some are expensive, some are free. Some are better for technical works, others do artistic or organic shapes better. Some are easy to learn, some take more practice. This book will use Blender since it is free and open source, has tools for modeling technical and organic shapes and is not too difficult to learn if you learn by doing.

Blender can be a bit tricky to get started with since it employs some conventions unique to its environment. Blender can be customized but this book will stick with the defaults so everyone is on the same page. Generally remember that *Ctrl + Z* undoes a multiple mistakes and can get Blender back to the state it was before, useful in tutorials to get back on track. The location of the mouse pointer is important when using Blender's hotkeys, which is the best way to learn to use Blender. Blender uses the right-click on mouse for selection by default. Finally, Blender's units translate to real life by 1 Blender grid space = 1 millimeter.

The next chapter will be a proper tutorial, teaching the most common modeling tools in Blender by inserting common shapes and manipulating them to the desired shape. Then that shape will be automatically smoothed to make it more appealing. Finally, the model will be edited and prepared for printing, utilizing the rules taught in this chapter.

2
Mini Mug

Blender has a variety of exciting tools for 3D modeling. This chapter will cover navigating the file system, adding objects, adjusting the view, selection tools, modification operators, object modifiers, and exporting an object ready for print. With so many things to learn on such a simple first project, by the time you finish you'll want to toast your success. So why not create something for that very purpose?

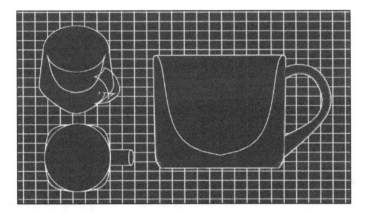

Our mug will be a miniature one, almost thimble-sized: 24 mm wide by 20 mm tall with a wall thickness of at least 2 mm. We'll put a handle on our mini mug to give it some character and give you something to hold on to. We'll also give it a little bit of body shape to make it more stable and printable. Beside from these we'll take advantage of its size to keep the details to a minimum.

Getting started

Blender opens with a default scene that contains a cube, a light, and a camera, none of which are necessary for this project. So to begin, the virtual work area will need to be cleared and a new file will be created for this project. Saving early is a good idea to name the project area. Saving often is a good idea in case anything bad happens; there is always a risk with any computer project. Saving incrementally is a good idea as a kind of back-up undo memory. And as this is the first project, a basic directory for the projects to follow will be set up. Carry out the following steps to create a new file in Blender:

1. Open Blender.
2. In the menu at the bottom of the 3D View panel choose **Select | (De)Select All** or press *A* to toggle the selection. As the cube was selected before this, the toggle will switch it so nothing is selected.
3. Choose **Select | (De)Select All** or press *A* again to select everything in the default scene.
4. Choose **Object | Delete** or press *X* to start the delete operation.
5. Click on **Delete** or press *Enter* to confirm the operation.

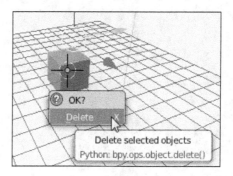

6. On the top menu click on **File** and choose **Save As....**

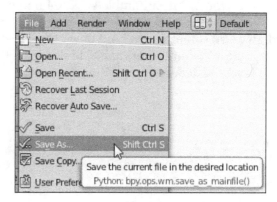

7. On the left-hand side bar under **System Bookmarks** click on **Documents**.

8. On to top, click on the **Create New Directory** button and click on **Create New Directory** in the menu that pops up.

9. Type in MakerbotBlueprints as the name for the new directory.

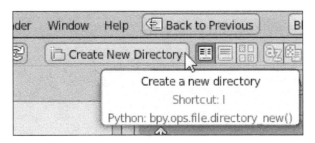

10. Click on the MakerbotBlueprints directory.

11. Click on the **Create New Directory** button and click again on the menu that comes up.

12. Type in Ch 2 MiniMug as the name for this directory.

13. Click on the Ch 2 MiniMug directory.

14. Click on the **untitled.blend** in the name bar and type in Mini Mug as the name of the project.

15. Click on the **Save As Blender File** button.

Creating the first shape

With this project's file created it's time to begin creating the mug. The mug will be made up of two different shapes, so the first thing to do is to add the shapes we need into the scene:

1. In the Info panel (top menu) click on **Add** or press *Shift + A* to add a new shape.

2. Mouse over **Mesh**.

3. Select **Cylinder** from the sub-menu that comes up.

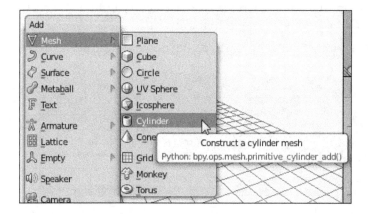

 Whenever you add something you have a chance to edit its options. Once these options are gone, you can't go back and change them. The new object is now just a collection of points, same as any other object to Blender so edit those options while you can.

4. On the left-hand sidebar under the **Add Cylinder** option, click on the number in **Vertices** and change that number to 8.

5. Press *Tab* to advance to the next option, **Radius**. The mug is designed to be 24 mm wide, so enter a radius of 12 and press *Enter*.

6. Press *Tab* to advance to **Depth**. Enter a depth of 20.

7. Choose **File | Save** or with the pointer over the 3D View panel press *Ctrl + S* and press *Enter* to save.

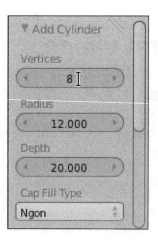

It is good practice to name the objects to avoid a bunch of nondescript objects such as "Cylinders" and "Cubes" in scenes. To name an object use the **Object** menu on the right-hand side bar and carry out the following steps:

1. On the right-hand side bar click on the icon that looks like an orange cube to switch to the **Object** tab.

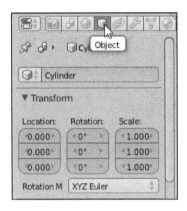

2. Click on the word **Cylinder** in the text box to select it.
3. Erase the word **Cylinder** and type Mug Body followed by *Enter* to name the object.
4. Select **File | Save** or with the pointer over the 3D View panel press *Ctrl + S* and then press *Enter* to save.

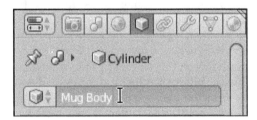

At this point the 3D view will turn into a field of gray. This is expected and will be addressed later after creating a save point.

Creating a save point

Incremental saves are when a new save file is created; leaving the old one in the state it was in. This serves as a sort of extended undo if anything goes wrong. In the case of following a tutorial like this these incremental saves can provide a way back if the reader ever goes astray and doesn't realize it at first, as opposed to having to start all over, avoiding frustration.

Blender contains a powerful shortcut to make incremental saving easy. Carry out the following steps for incremental saving:

1. Select **File | Save As...** or press *F2*.
2. With the pointer over the file list press + on the number pad to add a number to the file name.
3. Click on **Save As Blender File** to create the new file.

The next time this set of actions are followed the number in the file name will be incremented by 1 automatically.

Adjusting the view

In the main view there should be nothing but a gray field. This is because the mug is bigger than the current view so now is a good time to list the many ways to change the view in Blender.

- Rotate the view: Use the *8, 2, 4,* and *6* keys on the number pad to rotate the view. Use the *7, 1,* and *3* keys on the number pad to jump to the top, front, and right-hand side views. Use *Ctrl + Numpad 7, Ctrl + Numpad 1,* and *Ctrl + Numpad 3* to jump to the bottom, back, or left-hand side view. Pressing *5* on the number pad toggles orthographic/perspective view which means the view is either rendered with perspective like in real life or without like on a grid that is usually easier to do editing in. These options are also available in the **View** menu at the bottom of the 3D View panel. Click-and-hold the middle mouse button (or your mouse wheel) and move the mouse to rotate the view.

- Zoom the view: Use *Numpad +* and *Numpad –, Ctrl +* and *Ctrl –,* or spin the mouse scroll wheel to zoom in and out. Automatically center the view and zoom on all objects by pressing *Shift + C.* This also resets the 3D cursor to the origin. Center the view and zoom on the selected objects by pressing the .(period) key on the number pad.

- Pan the view: *Shift* + click-and-hold the middle mouse button to pan the view, or in other words, to move the view without changing the direction you're looking at the scene. Alternately you can press *Shift + F* to begin the "Fly Camera" operation. In Fly Camera move the mouse pointer towards the edge of the screen to turn the camera. The mouse wheel or + and – on the number pad will let you move forward or backwards. *Left-click* or press *Enter* to exit the Fly Camera operation and leave your view where it is at the end. *Right-click* or press *Esc* to cancel the Fly Camera operation and return to where you started.

Using the view commands adjust the view. In future projects, there will be less emphasis on the specific keys to press to change the view, but for now the steps to carry out are as follows:

1. In the menu at the bottom of the 3D View panel choose **View | Align View | Center Cursor and View All** or press *Shift + C* to center the view on the cylinder.
2. Select **View | Front** or press *Numpad 1* to switch to front view.
3. Select **View | Orthographic** or **Perspective** or press *Numpad 5* to switch between them.

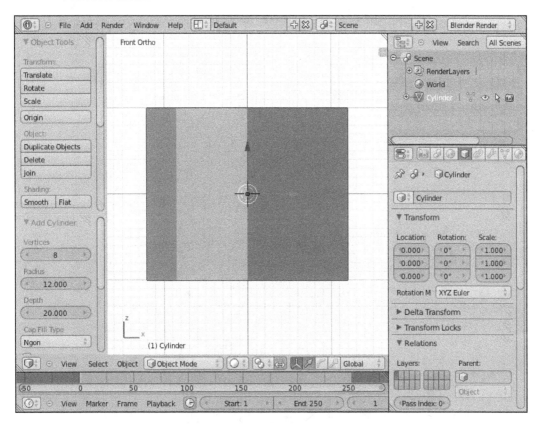

4. In the bar at the bottom of the 3D View panel click on the combo box with an icon like a white ball and choose the **Wireframe** option or press *Z* to switch to wireframe view.

 In **Wireframe** view the object is presented as though it were just made of wire. All edges, even hidden ones, become visible.

Adding a handle

Do an incremental save (Press *F2*, *Numpad +*, and then click on **Save As Blender File**).

 Whenever a new object is added, it will appear wherever the 3D cursor is. If the 3D cursor has accidentally moved by a stray, left-click is good to center the view and reset the 3D cursor to the origin by pressing *Shift + C* before adding a new object.

Carry out the following steps to create a new cube:

1. Select **Add** in the Info panel menu or press *Shift + A* and select **Mesh | Cube** in the menu.

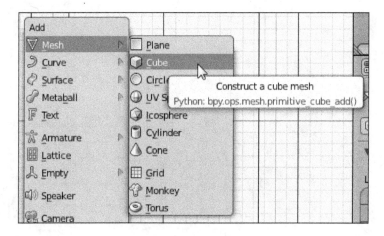

2. Click on the **Object** tab in the right-hand sidebar and rename this **Cube** to Handle.

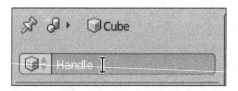

The newly created cube is being seen from the front orthographic view so it looks like a box, but it does have depth. Rotate the view to confirm this. The cylinder is also entirely inside the mug body but it is visible because of wireframe mode. Press Z to toggle back to solid view and the cube will be hidden by the body. Remember to undo any view changes by pressing *Numpad 1* or choosing **View | Front** and toggling to **Wireframe** mode (Z) before continuing. The steps for the grab operation are as follows:

1. From the menu at the bottom of the **3D View** panel choose **Object | Transform | Grab/Move** or move the mouse pointer over the handle cube and press G to begin the grab operation.

2. Move the mouse until the handle is inside the lower-right corner of the mug body.

3. Press *Enter* or *left-click* to end the grab operation.

4. Press *Ctrl + S* and *Enter* to save.

Object manipulation such as movement, rotation, or scaling are all done by default relative to the current view. As the current view is the front view, the grab operation will only move up and down or left and right, or along the x and z axes.

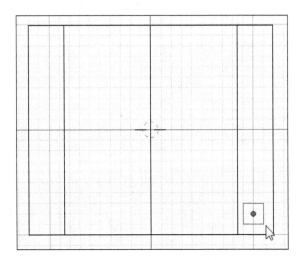

Carry out the following steps for the rotate operation:

1. Choose **Object | Transform | Rotate** or press R to begin the rotation operation.

2. Type in 45 to rotate the handle.

3. Press *Enter* or *left-click* to end the rotate operation.

4. Press *Ctrl + S* and *Enter* to save.

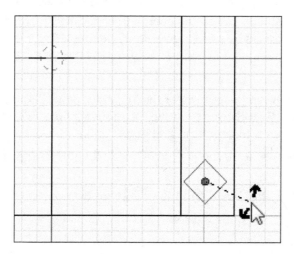

Shaping the handle

Pan and zoom the view as explained before to adjust the view to focus on the right half of the mug with plenty of space to make the handle in.

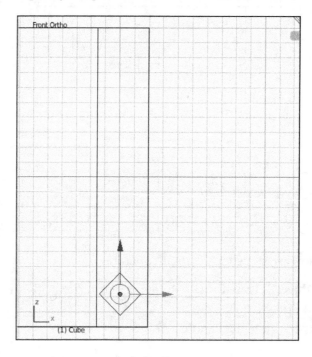

Blender allows direct manipulation of the individual points or vertices that make up an object in **Edit Mode**. Many things that you can do in terms of selection and manipulation work similarly in and out of **Edit Mode**.

1. On the menu at the bottom of the 3D View panel click on the combo box with the option for **Object Mode** visible and select **Edit Mode** or press *Tab* to enter **Edit Mode**.

2. Choose **Select | (De)select all** or press *A* so that no points are selected.

3. Choose **Select | Circle Select** or press *C* to begin the circle select operation.

4. Scroll the mouse wheel to adjust the circle select tool size.

5. Hold left mouse button and to move the selection area to select the vertices shown in the next screenshot.

6. *Right-click* or press *Enter* to end the circle and select operation.

In **Wireframe** mode selecting the vertices one at a time (with the *right-click* on mouse) cannot be done in confidence as the vertex selected may be any of the overlapping vertices, if overlapping vertices there are, as in this case. Using the circle select operation selects them all.

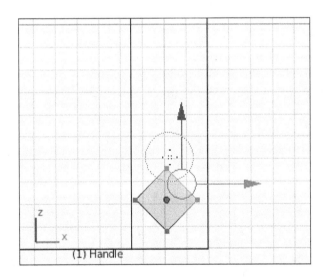

(1) Handle

The extrude operation creates a new points based on the selected points and allows you to extend or move these shapes away from where they started. This is a quick and easy way to change the geometry of the shape.

1. Select **Mesh | Extrude Region** or press *E* to begin the extrude operation.

2. Move the mouse to move the extruded selection away from where they started.

3. *Left-click* or press *Enter* to end the extrude operation.

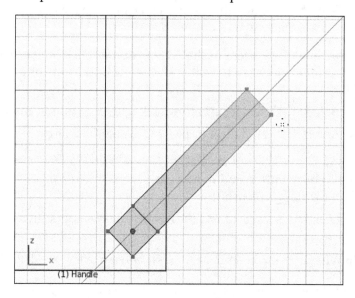

4. Select **Mesh | Transform | Rotate** or press *R* to begin the rotation operation.
5. *Left-click* or press *Enter* to end the rotation operation.
6. Select **Mesh | Transform | Grab/Move** or press *G* to begin the grab operation.
7. Use the mouse to set the location similar to the illustration.
8. *Left-click* or press *Enter* to end the grab operation.
9. Move the mouse to rotate the points similar to the illustration in the following screenshot:

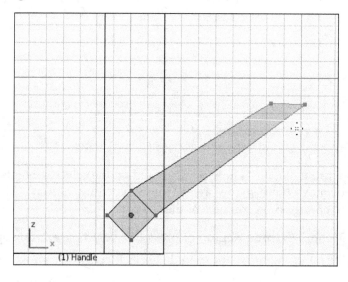

Building the rest of the handle is as easy as; **Extrude** (*E*), **Rotate** (*R*), **Grab/Move** (*G*), repeat. At each stage be sure that the points you created approximately match what you see in the illustration in the following screenshot. Keep it rough at this point. Smoothing out the mesh comes later.

Make sure to remain in front view (*Numpad 1*) when doing the operations or the handle may not remain straight in space. If that occurs remember repeatedly selecting **Mesh | Undo** or pressing *Ctrl + Z* undoes many mistakes. In the worst case reload the last incremental save and back up to that point.

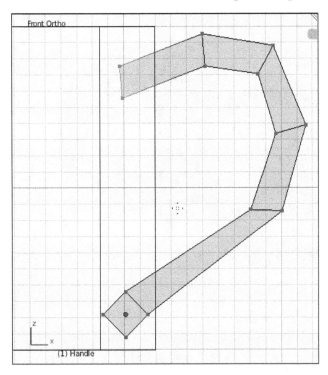

Once the rough shape of the handle is complete, exit **Edit Mode** either by selecting **Object Mode** from the combo box at the bottom of the **3D View** panel, or by pressing the *Tab* key. Then either choose **Solid** from the display combo box or press *Z* to switch to solid view, adjust your view to see how the handle looks in three dimensions.

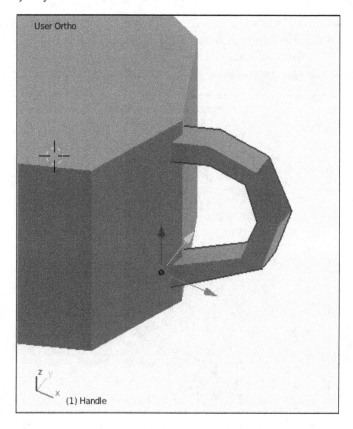

Before proceeding do another incremental save (press *F2*, then *Numpad +*, and click on **Save As Blender File**).

The handle is in the right shape, but too thin. This is what the scale operator is for. Carry out the following steps for scale operation:

1. With the handle selected, navigate to **Object | Tranform | Scale** or press *S* to begin the scale operation.

2. Move the mouse and you will notice that the whole handle grows bigger and smaller.

3. Press *Y* to lock the scale to only the y axis.

4. Move the mouse and notice the handle is only scaling in the y axis.

5. Move the mouse until the handle is about three times thicker than it was or type 3. There's no text box to type this into. Simply typing during the operation will define the parameters of the operation in Blender.

6. Press *Enter* or *left-click* to end the scale operation.

7. Save (*Ctrl + S*).

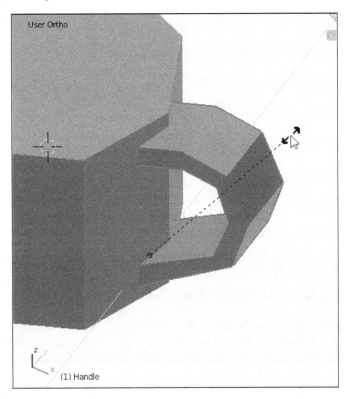

Smoothing the rough edges

Blender provides a number of object modifiers that can be used to quickly alter objects in the scene. In this project only two modifiers, namely **Multiresolution** and **Boolean**, will be used. Modifiers can be stacked, turned on and off, and their settings can be changed on the fly. The modifiers are accessed in their own menu in the right-hand side bar.

Begin with an incremental save (Press *F2*, then *Numpad +*, and click on **Save As Blender File**) to set a save point. Then carry out the following steps to create a new modifier:

1. In the right-hand side bar, click on the icon that looks like a wrench to open up the modifier tab.

2. Click on the **Add Modifier** button.
3. Select from the second column (**Generate**) the **Multiresolution** modifier from the menu.
4. In the **Multiresolution** options box click on the **Subdivide** button twice.

Multiresolution smoothes the mesh out by adding more vertices between existing vertices and putting them in a location that rounds the shape. In this case the final result looks good except that the top of the handle is too rounded so it doesn't join the body of the mug well. To flatten these curves out more points will need to be added to the original mesh. Fortunately modifiers do not change the original geometry until you click on the **Apply** button, so you can still modify the shape in **Edit Mode**.

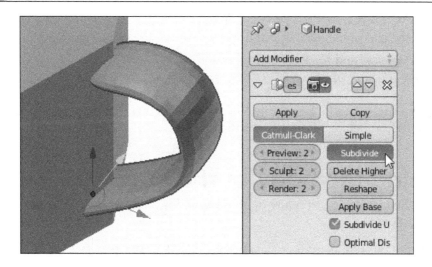

To further modify the shape carry out the following steps:

1. Enter the **Wireframe** view (*Z*).

2. Enter **Edit mode** (*Tab*).

3. The **Loop Cut and Slide** operation is accessed either by finding the button in the left-hand side bar under **Add** or by pressing *Ctrl* + *R*.

4. Move the pointer with the mouse to the lines near the top at the end of the handle, as shown in the next screenshot.

5. *Left-click* to select this subdivision.

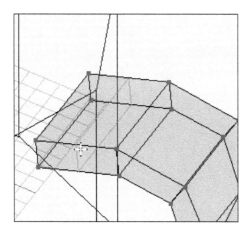

6. Move the pointer closer to the end of the handle, to slide the loop cut closer to the end of the handle, but not overlapping the existing points.

7. *Left-click* to end the loop-cut operation.

8. Exit **Edit Mode** (*Tab*).

The top end of the handle is now blunter and will sit inside the mug better. Save the work done so far before continuing further.

Shaping the body of the mug

Finishing the handle taught many operations and modifiers that will now be applied to the mug body:

1. Begin by setting another save point (Press *F2*, then *Numpad +*, and click on **Save As Blender File**).

2. *Right-click* to select the **mug body** or choose the **mug body** in the Outliner panel.

3. Jump to the front view (*Numpad 1*) and center and zoom (*Numpad .*) on the mug body.

The border select operation is another way to quickly select multiple objects or, in this case, points for editing. Like the circle select tool while in wireframe mode, the border select tool selects all vertices that it surrounds, even if they're overlapping other points on the screen. The following are the steps for border select operation:

1. Enter **Edit Mode** (*Tab*).

2. Clear the selection (*A*).

3. **Select | Border Select** or press *B* to begin the border select operation.

4. Hold left-button on mouse and move the pointer to draw a box around the vertices at the top of the mug body.

5. Release the left-button on mouse to end the box select operation.

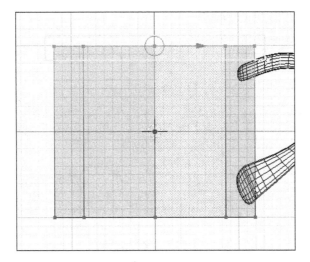

With the top of the mug body selected, press **Scale** (*S*) the selected points. Make the top of the mug slightly wider than the bottom. By default the center of the scale operation is relative to the points selected. In this case that works out so the top of the mug widens evenly all around. (This can be a problem for less regular shapes.)

 As with most operations, where the mouse is at the start of the scale operation, can affect the outcome. If the mouse pointer is too close to the center, there may not be a high enough degree of control, too close to the edge of the screen and there may not be enough room to move outward. This idea of pointer placement before an operation may take some getting used to but is a powerful reason to use the hot keys because that is not possible with menu select operations. If the pointer placement is ever undesirable simply press *Esc* to cancel the operation and try again.

Do a 1 millimeter extrusion (*E*) of the points at the top of the mug by typing 1 during the extrude operation. The next operation will use extrude in a different way to create points for scaling.

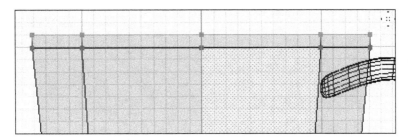

Following are the steps for extrude operation:

1. Do not touch the mouse.
2. Begin the **Extrude** (*E*) operation.
3. Do not move the mouse.
4. Press *Enter* to end the extrude operation.

It is not immediately apparent but the newly extruded points are exactly in the same place as the points they were copied from. Generally having duplicate points like this isn't a good thing, but the new points aren't going to stay where they are.

Pan (Press *Shift + middle-click* on mouse) and zoom (using mouse wheel) the view to look closely at one of the upper corners of the mug body. Put the mouse pointer close to the corner of the selected points and **Scale** (*S*) them until they are approximately 2 mm inside the mug body.

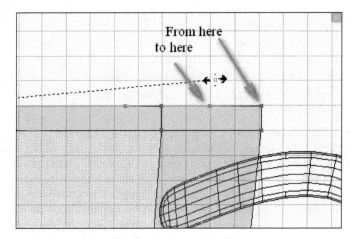

By adjusting the view, the result of this operation becomes apparent. This is how the lip of the mug is created.

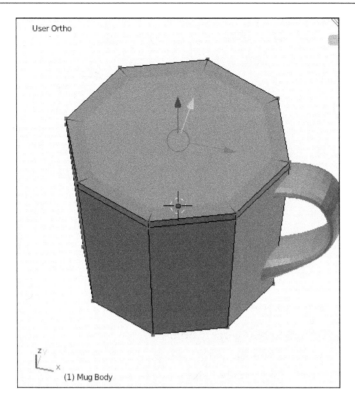

Jump back to the **Front** view (*Numpad 1*) and adjust the view to include the whole mug body. **Extrude** (*E*) again, but this time move the extrusion downwards into the mug body until the extrusion is nearly 2 mm, or 2 grid blocks, from the bottom of the mug on the inside. Then save (*Ctrl + S*).

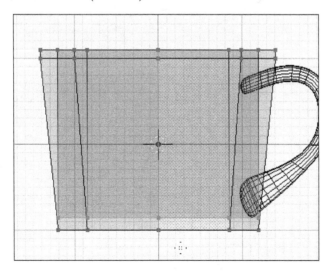

Zoom (using mouse wheel) and pan (Press *Shift + middle-click*) to adjust the view to look closely at a lower corner of the mug to fine tune the last extrusion operation.

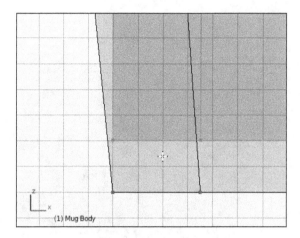

Carry out the following steps for the grab operation:

1. Begin the **Grab/Move** (*G*) operation.
2. Lock movement to the z axis by pressing *Z*.
3. Move the mouse until the bottom is closer to exactly 2 grid blocks (2 mm) from the bottom.
4. **Scale** (*S*) until the bottom of the mug is approximately 2 mm inside the wall of the mug.
5. Save (*Ctrl + S*).

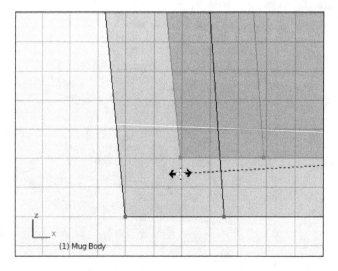

On a Makerbot it is possible to make things with details as small as 0.4 mm thick because that is the size of the nozzle. However, such small things tend to break easily. Even details as much as 1 mm can be too fragile. The general rule to follow is anything you don't want to break needs to be at least 3 mm thick or more and walls should never be made thinner than 2 mm.

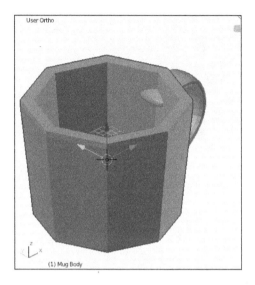

The mug is looking very mug-like now, but it's still very rough and the handle is peeking through the body. First thing to do is in the right-hand side bar switch to the modify tab (the one with the wrench) and hit the **Add Modifier** to add the **MultiResolution** modifier to the mug body.

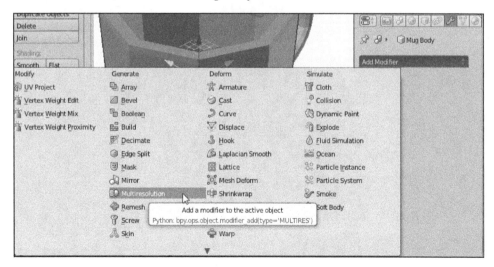

Click on the **Subdivide** button three times. Now the mug body looks more like a cup, but it doesn't have a stable base for printing. This is an excellent opportunity to give the mug an interesting shape.

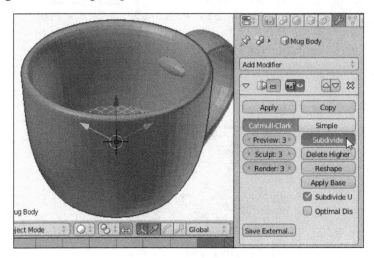

Carry out the following steps to give an interesting shape to the mug:

1. Enter **Edit Mode** (*Tab*).
2. Clear the selection (*A*).
3. Jump to **Bottom** view (*Ctrl + 7*).

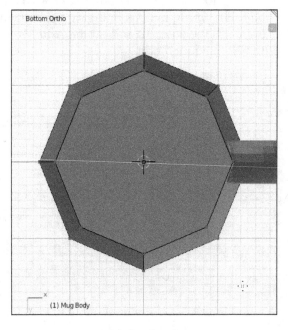

4. *Right-click* to select one of the vertices on the bottom of the mug.

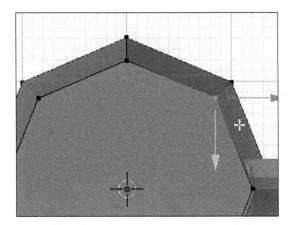

5. While pressing the *Shift* key, *right-click* on every other vertex on the bottom of the mug body.

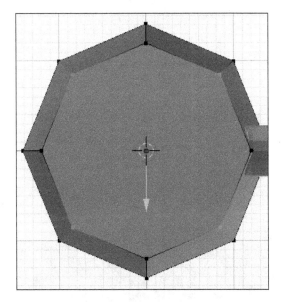

6. Move the mouse pointer close to any of the selected points.

7. Begin the *Scale* (*S*) operation.

8. Move the mouse pointer away from the middle until the bottom of the mug body takes on a more-or-less square shape.

9. Press *Enter* to end the scale operation.

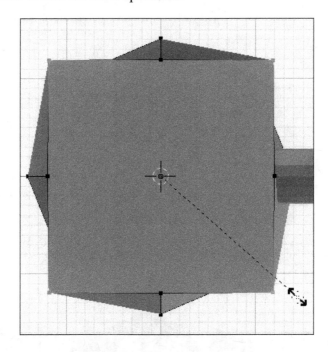

If you exit **Edit Mode** at this point the **Multiresolution** modifier will still be making the bottom of the mug too rounded. This is the same as what happened with the handle and is fixed in the same way.

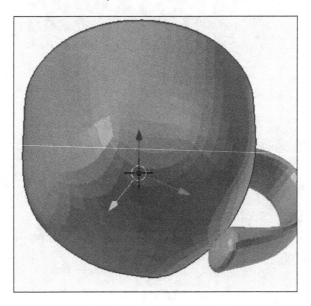

Rotate the view upwards slightly to be able to see the side of the mug. Use the loop cut operator (*Ctrl + R*) to add extra points around the bottom of the mug near the end. Remember with the loop cut operator the left-selects the edges to be cut, then move the mouse to adjust the location of the cut, then *left-click* to again set the loop cut.

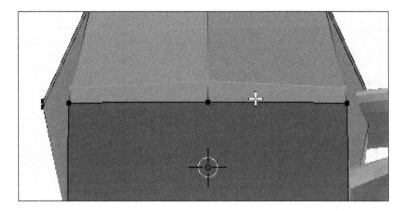

Exit **Edit Mode** (*Tab*) and adjust the view. The mug now has an interesting shape and more importantly a flatter base.

It is time to fix the handle. This is a good time to set a save point (Press *F2*, then *Numpad +*, and click on **Save As Blender File**). Select the handle jump to the front view (*Numpad 1*) and pan (*Shift + middle-click*) the view. Enter **Edit Mode** (*Tab*) and in **Wireframe** view (*Z*) select the points on the top of the handle that are extending too far with either the **Border** (*B*) or **Circle** (*C*) select tools.

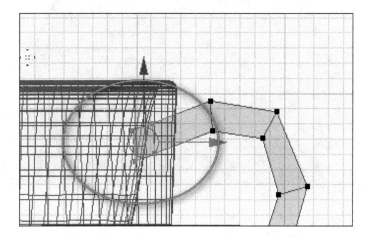

Move the points with the **Grab/Move** operator (*G*) until they are inside the wall of the mug and exit **Edit Mode** to see if the handle stays within the walls of the mug after the modifier is applied. Then save (*Ctrl + S*).

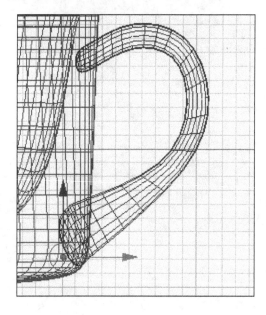

Joining the shapes

Most of the modeling is done, but this mug is not quite ready to print. The body and handle are still two separate pieces and the bottom of the body, while flatter, isn't quite flat enough to trust to printing. We will be making some changes that will make further editing difficult so first we will make a duplicate of the mug's body easily modifiable objects.

Even though it's only been a while set another save point (Press *F2*, then *Numpad +*, and click on **Save As Blender File**) now.

Carry out the following steps to make a duplicate of the mug:

1. Center the view on all visible objects (*Shift + C*).
2. Select the mug body (*right-click*).
3. Do not touch the mouse.
4. **Object | Duplicate** or press *Shift + D* to begin the duplicate operation.
5. If the mouse is moved the duplicate will move which can be useful at times but do not move the mouse this time.
6. Press *Enter* to end the duplication operation.

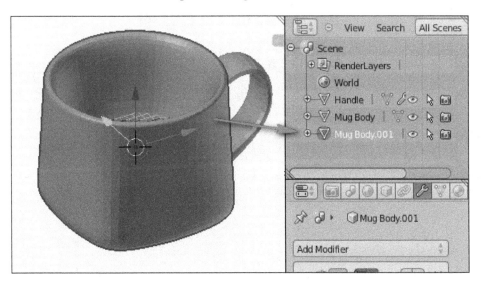

With two identical shapes in the same location it can get confusing to select the correct one, so we're going to hide everything but the duplicate object that we're working with. The objects will still be in the scene there, but they'll be invisible and not selectable. In the menu select **Object | Show/Hide | Hide UnSelected** or press *Shift + H* to hide all objects but the selected one.

Use the **Object** tab (with the orange cube) to change the name of **Mug Body.001** to
Mug Final. Remember to save.

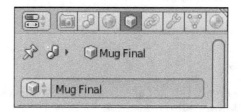

The Mug Final will be the version that will be prepped for printing. The body of
the handle will be merged into this version before some minor edits to the mesh will
make it ready to print. To begin, click on the **Mesh** tab in the right-hand side bar
(the one with the wrench) and click on **Apply**.

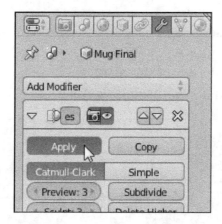

 After applying the modifier, if you go to **Edit Mode** you will notice
that we can now only modify the high resolution mesh. This is good
to do to prepare the mesh for printing but as it cuts off some editing
options so it is best to save this step until the end.

Now a new modifier, **Boolean**, will be used to put the handle on the mug. The two objects will have their geometry combined and any geometry that would be "inside" the final object will be eliminated:

1. Click on the **Add Modifier** button.
2. Select the **Boolean** modifier.

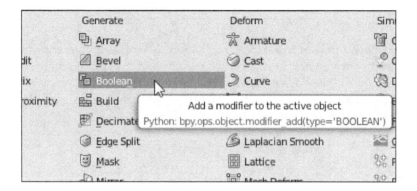

3. In the **Boolean** options click on the button below the word **Operation** which reads **Intersect**.
4. Choose **Union** from the menu that pops up.

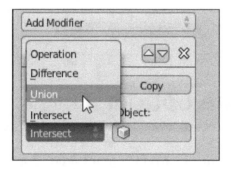

5. Click on the box under the word **Object**.
6. Choose **Handle** from the menu that pops up.

7. Click on the **Apply** button to finalize the **Boolean** modifier.

The handle is now a part of the Mug Final object. If you switch to wireframe view (Z) and change your views you will notice the parts of the handle that were inside the mug aren't there. They are joined so that the whole shape is one continuous shell without hidden vertices inside the shape. This is a best practice for making objects for 3D printing. Remember to save.

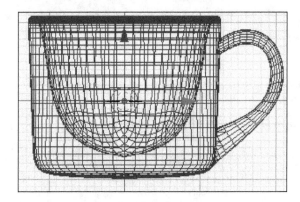

Flattening the bottom

The bottom of the mug, while flatter than before, is not perfectly flat and is therefore not suitable for printing yet. However, there is a way to use the scale operator in **Edit Mode** to make a flat bottom, by carrying out the following steps:

1. Jump to the front view (*Numpad 1*).
2. Enter **Edit mode** (*Tab*).
3. Clear the selection (*A*).

4. Use the **Box Select** operation (*B*) to select the bottom few layers.

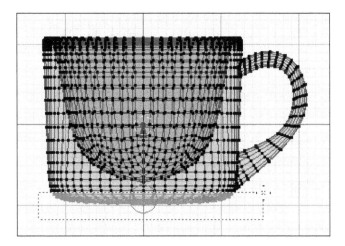

Without touching the mouse carry out the following steps:

1. Begin the **Scale** (*S*) operation.
2. Press *Z* to lock the scale operation to the z-axis.
3. Type 0 (zero) to scale by a factor of zero.
4. Press *Enter* to end the scale operation.

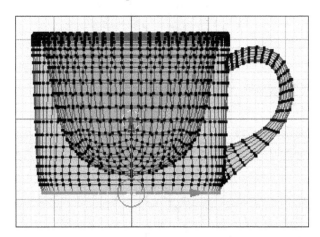

Rotate the view and notice that all the points that were there before are still there, they are all just on the exact same z level making a perfectly flat bottom for printing.

Remember to save (*Ctrl + S*).

Exporting for print

All editing of a model done. The mug isn't placed on the platform it's true, it is just kind of floating in space, but it is properly oriented so the 3D printer software will take care of positioning it. All that is left is to export the model in a format that can be sent to the 3D printer.

With the final mug selected (*right-click*) and when *not* in **Edit Mode** (*Tab*), carry out the following steps:

1. Click on **File** on the top menu.

2. Click on **Export** on the menu.

3. In the menu that pops up click on **Stl (.stl)**.

 Standard Tessellation Language (STL) is a file format that is used to describe the geometry of a shape. It doesn't store much else but the shape which is perfect for 3D printing. Almost every 3D printer can use a STL file.

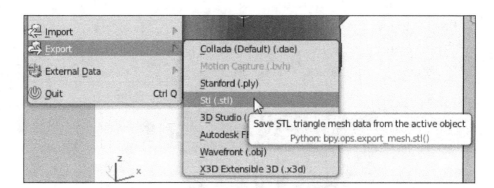

By default the name of the STL will be the name of the project, which is perfect in this case. Just check to be sure the name is as you like it and click on the **Export STL** button.

Then open either MakerWare or ReplicatorG, open the STL, and prepare it for printing in the usual way.

Extra credit

Now that you've learned the basic modeling tools challenge yourself to make your own mug shape, either by modifying the existing model or starting from scratch. Make a mug that is more of a goblet or make a mug with extruded tentacles. Perhaps, something that is more irregular and less symmetrical. The possibilities are endless.

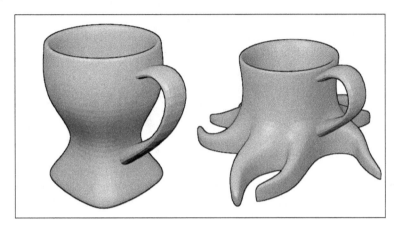

Summary

And that's it. This thimble-sized mug taught a majority of the basic tools that will continue to be used in future 3D modeling projects. Some of those tools are:

- File operations such as saving and creating new directories
- Adding basic objects (*Shift + A*)
- View rotation with the number pad
- Scene navigation with the middle-button on the mouse
- **Wireframe** and **Solid** view modes (*Z*)
- Selection operations such as **Circle** (*C*) and **Box** (*B*) select
- Manipulation operations such as **Grab/Move** (*G*), **Rotate** (*R*), and **Scale** (*S*)
- **Edit mode** (*Tab*) for manipulation of the individual elements that make up a shape
- The **Extrude** (*E*) operation
- The loop-cut (*Ctrl + R*) operation
- Object modifiers such as **Multiresolution** and **Boolean** operations
- Exporting STLs for print

That is a long list. Have a drink.

The next chapter will teach a different tool set for modeling similar objects using lines and lathes, but the skills of vertex selection and manipulation learned in this chapter will still come in useful!

3
Face Illusion Vase

There is an old illusion where the silhouette of two faces are shown towards each other with their noses close together, but if you look at the space between the faces, the image shifts in your eyes and appears to be of a single ornate vase. With 3D printing, it is possible to make this illusion a reality. More than just a reality, it is possible to use a real face to make the illusion.

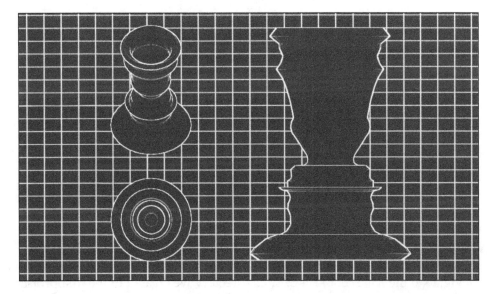

In the previous chapter, a number of topics including manipulating the view, and creating and manipulating objects were introduced. Also, the habit of making frequent saves and using incremental saves were taught, practiced, and enforced. In this chapter and in all the following chapters, saving will be left to you. It's generally a good idea to make an incremental save at every major section marked by a heading, and doing smaller saves as much as possible.

This chapter will focus on importing an image to help guide modeling, the tools that can be used to trace that image with a single line, and the tools that can be used to turn a single line into a solid shape.

Getting a profile

Before opening Blender, the first thing to do is get a side profile image of a face. With a camera you can take an image of yourself or someone you know. Alternately, searching online for side profile picture can turn up something usable. A good picture will be as side-on as possible and will have a clear outline of the face. Save the image to the local disk and make note of its location. If no other image can be found, the following image can be acquired at `http://www.thingiverse.com/ thing:90754`. Locate the following image, right-click on it and choose **Save Image As...** (or **Save As...** depending on the browser) and save the image in a directory, where it can be found later:

Now, just like last time, open up Blender and clear the scene. Start a new project, **Select All** objects (*A*) and **Delete** (*X*). In the menu bar, navigate to **File | Save As...** and navigate to the `MakerbotBlueprints` directory we created in the first project, then create a new directory for this project and in that directory name this file `Face Vase`.

Importing an image for reference in Blender is done through the **Properties** menu, a special menu which is usually hidden. First that menu will be revealed and then the image will be imported to be used as reference:

1. Open the **Properties** menu either by selecting from the bottom menu of the 3D view's window **View | Properties** or by pressing *N*:

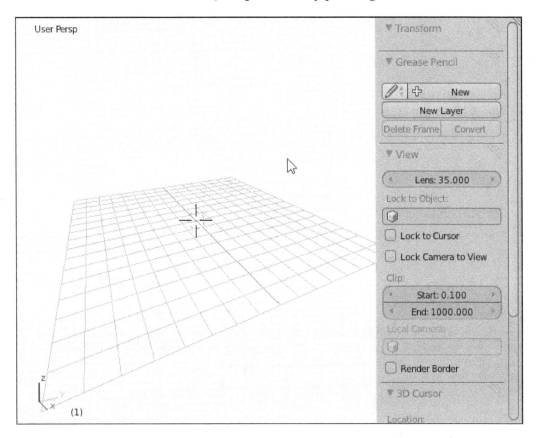

2. Scroll down the **Properties** menu and check the checkbox next to the **Background Images** submenu.

3. Expand the **Background Images** section by clicking on the arrow next to the checkbox.

4. Click on the **Add Image** button:

5. In the options box that opens up, click on the **Axis** combobox that currently reads **All Views** and click on **Front**:

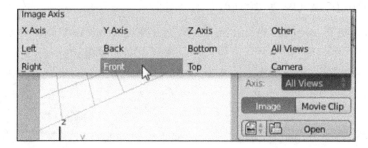

6. Click on the **Open** button:

7. Navigate to where the profile image is saved.
8. Double-click on the image or select the image and click on the **Open Image** button.

The image will not be immediately visible because it is only set to appear in the front view so switch to the **Front** view (*Numpad 1*) and switch to **Orthographic** view (*Numpad 5*) and the selected image will appear:

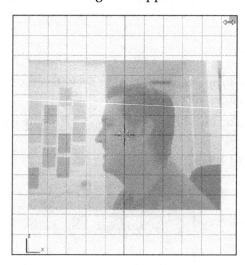

Now in the **Properties** menu under **Background Images**, this image will have some new options to manipulate the image. This project is more easily done with the opacity set near to or at 100 percent, making a very dark and clear image, so raise the **Opacity** option by either clicking and dragging on the slider or clicking on it and typing the value you want. (Remember that saving is always a good idea.):

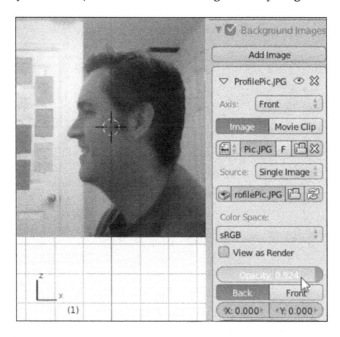

Tracing the silhouette

In Blender, an object can be just a single-shaped line that can later be turned into a 3D object. Blender doesn't have a single-line object, so instead the simple objects if possible will be added and a single line will be extracted from it. Perform the following steps to trace the silhouette:

1. Add (*Shift + A*) an object.

2. Under the **Mesh** menu, click on **Plane**:

3. Begin the **Rotation** (*R*) operation.

4. Press *X* to lock the rotation around the x axis.

5. Type 90 to rotate exactly 90 degrees.

6. Press *Enter* or *left-click* to end the rotation operation.

7. In the **Object** tab (the one that looks like an orange cube) on the left-side bar, name this object Face Line.

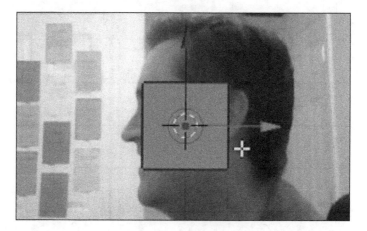

Unlike before, when we added a cube that only looked like a square because of our viewing angle, a plane really is just a square and has no depth. A plane isn't enough to define a printable shape but it can be used as a starting point for making a printable shape with depth.

All 3D shapes are comprised of some basic building block. **Vertices** (single vertex) are the smallest part and are nothing more than a point, no shape or depth, and are not printable on their own. Two vertices can be connected with a **line**. Lines also aren't printable on their own. When three vertices are connected with lines the result is a **face**, sometimes called a polygon in other applications. Faces are 2 dimensional objects. In Blender, there are **fGons**, polygons with more than three vertices and lines. Blender often treats these fGons just like faces, but really they are collections of polygons that all act together. The plane is nothing more than a four-sided fGon sometimes called a quadrangle.

Finally, with a collection of faces a 3D shape can be created. Shapes are 3-dimensional objects and may be printable.

With the plane selected, a single point will be isolated from it and that point extruded into a line to trace the face with:

1. Enter the **Edit Mode** (*Tab*).

2. Select one of the vertices on the right side of the plane with the *right-click*.

3. Press and hold *Shift* while selecting the other vertex on the right side of the square with the *right-click*.

4. **Delete** (*X*) the selected vertices:

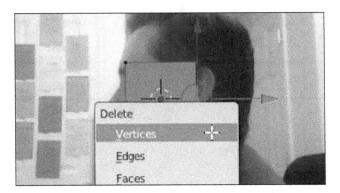

The remaining line will be the starting point for the traced silhouette.

5. Select (*right-click*) either remaining vertex.

6. **Grab/Move** (*G*) and move it to a point on the edge of the face.

7. Select (*right-click*) the other point and **Grab/Move** (*G*) and move it to a different point on the edge of the face:

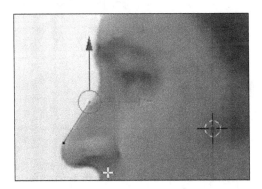

8. Select (*right-click*) the top vertex and **Extrude** (*E*) to create a new point. Place it on a point higher on the edge of the face:

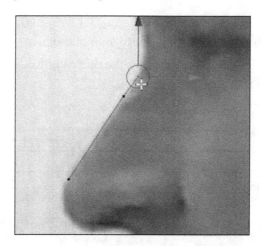

9. Repeat this process as many times as necessary to trace the top of the face. Zoom and pan as necessary:

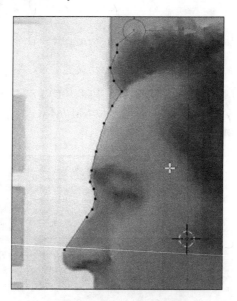

Remember the rules about overhang. Do not go over the top of the head. Stop before the overhang become too steep; about 45 degrees should be the max. It is not necessary to place the points at exactly the same distance from each other. As a general rule, the fewest points necessary to catch all the details is best.

Now continue tracing the bottom of the face:

10. Select (*right-click*) the lowest vertex and repeat the extrusion and placement process to trace the bottom of the face:

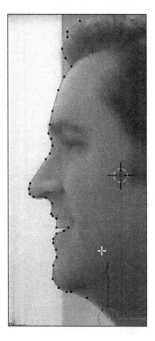

11. Exit the **Edit Mode** (*Tab*).

12. Turn down the opacity in the **Properties** panel on the **Background Image** options to inspect the traced silhouette:

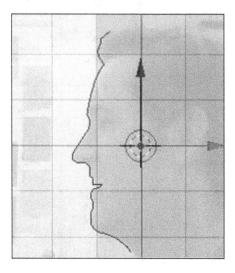

If it is necessary, enter the **Edit Mode** again and move points around until they are where you want them. If you need to add a point between two points don't use extrude. Instead, select the two points you want to add a point between and click on the **Subdivide** button in the **Object Tools** section on the right side bar.

If the trace looks good, open the **Properties** menu again and under the **Background Images** options click on the icon shaped like an eye near the top to hide the background image entirely:

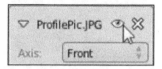

Creating a vase from the lines

It may not be immediately evident from its name, but the **Screw** modifier is the tool that will be used to turn this line into a 3-dimensional shape:

1. Switch to the Modifier tab (the one with the blue wrench) in the right-side bar.
2. Click on the **Add Modifier** button.
3. Click on the **Screw** modifier in the second column:

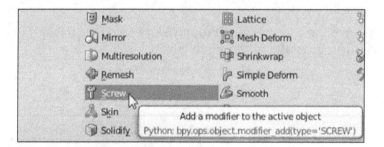

4. Uncheck **Smooth Shading**.
5. Change the **Render Steps** field to **32**.
6. Change the **Steps** field to **32**:

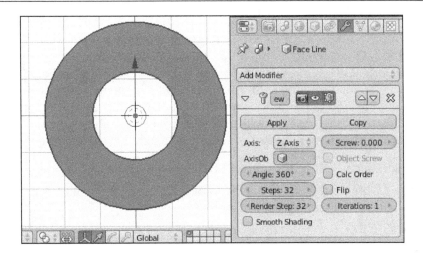

Like many modifiers in Blender, **Screw** can be used for many more things than making vases, and experimentation outside of this exercise is encouraged.

Why 32 steps? Those who are new to 3D modeling or computer science may wonder why such a seemingly arbitrary number like 32 is used. There are two reasons for choosing 32.

The first reason is because 32 is divisible by 4, so that the vertices line up with the four axes directions nicely. If 30 were used, there would be vertices that lined up nicely with the horizontal axis, but on the vertical axis there would be points straddling it, which might be complicated if you have to do any vertex editing along the axis later. It's not a hard-and-fast rule, but generally speaking it's best to choose numbers which are divisible by 4 and have small prime roots, hence 8, 16, 32, and 64 are common.

The second reason is because 32 is sufficiently high to make the shape seem round. For smaller objects, 16 may be enough and generally it's good not to overdo it. In the following screenshot, the effect of 8, 16, 24, and 32 steps are shown. By 24 the object is pretty round, and 24 is a pretty good number of steps to choose. But for the reason given before, 32 is better and not that much more. However, there is a point of diminishing returns and 64 steps (not shown) has hardly any visible benefit.

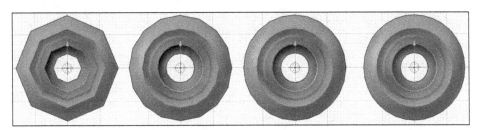

The **Screw** modifier options indicate that the rotation should be happening around the z axis, but this rotation appears to be happening around the y axis instead. This is because the rotation is happening around the object's z axis, not the world's z axis. Objects have orientations independent of the world's orientation and often modifiers use an object's orientation for their transformations. The plane that this object started as was rotated 90 degrees around the x axis at the beginning, so it's personal z axis is now oriented along the world's y axis. It is sometimes desirable, like in this case, to apply an object's rotation to fix it's orientation to match the world's orientation:

1. Be sure Blender is in the **Object Mode** (*Tab*), not the **Edit Mode**.

2. In the **Object Mode** from the menu at the bottom of the **3D View** panel navigate to **Object | Apply** or press *Ctrl + A*.

3. Click on **Rotation** from the menu:

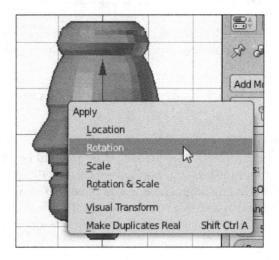

Now the object's orientation matches the world's orientation and the rotation automatically spins around the z axis as desired.

But the rotation is still not quite right. The outline is rotating around the middle with the nose out, when it should be nose in. The easiest thing to do is move all the points until they rotate in the desired way:

1. Enter the **Edit Mode** (*Tab*).

2. Select all the points (*A*).

3. **Grab/Move** (*G*) and move them to the other side of the z axis.

4. While in the **Edit Mode**, zoom out and **Scale** (*S*) the vertices until the vase is approximately 40 mm tall.

5. If necessary move (*G*) the points again to keep the thinnest point of the vase, where the noses are, about 10 mm thick.

6. When satisfied with the shape of the vase, exit the **Edit Mode** (*Tab*).

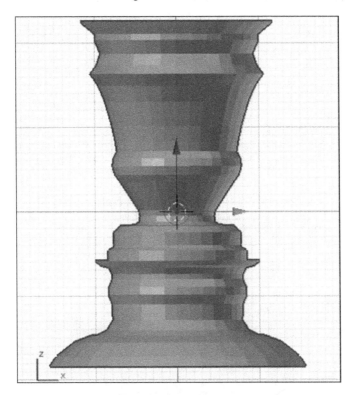

Using Solidify to make walls

The vase looks good from this angle, but if the view is rotated it becomes apparent that this it is just the shell of a vase. If this were to be printed as it is, nothing might get printed. The shape needs to be solidified into a shape, and fortunately there's a modifier for that. **Solidify** can be applied to this shell to make a printable object.

Just a quick reminder that constantly saving (*Ctrl* + *A*) and incremental saves (*F2*, +) is always a good idea.

Before using **Solidify** in the Modifier tab (blue wrench), press the **Apply** button to finalize the **Screw** modifier. Then add and adjust the settings of the **Solidify** modifier as follows:

1. Press the **Add Modifier** button.

2. Select the **Solidify** modifier in the second column from the pop-up menu:

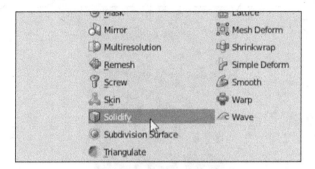

3. Change the **Thickness** field to **2** for a 2 mm wall:

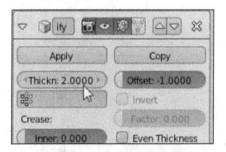

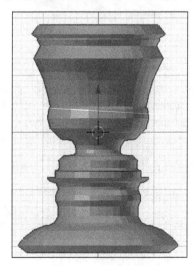

There is something very wrong. The **Solidify** modifier seems to work properly for the bottom half of the vase, but not for the top. The reason for this is another very important concept in 3D modeling; **Normals**.

 A collection of vertices and edges makes a face as discussed. But there is also another very important piece of information contained in a face, which direction it is facing or the face's normal. In the past, this information was used to inform the computer when they could ignore a polygon. If a polygon is facing away from the view, then the computer could save a few cycles and not draw that polygon. Now, computers are more powerful so polygons are drawn no matter which direction it is facing, but the normal is still used in 3D modeling to assist modifiers and operations. The normal is used by the extrude operation and now by the **Solidify** modifier.

The normal on the vase needs to be inspected and corrected before the **Solidify** modifier will work properly:

1. Enter the **Edit Mode** (*Tab*).

2. If the **Properties** panel was closed, press *N* to open it.

3. Under the **Mesh** display, find the **Normals** section and click on the Display Face normal button, which looks like a cube with one orange face:

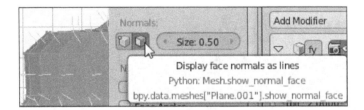

Part of the vase is now covered with visible blue hairs, and the other part, the trouble area, is hidden behind the effects of the **Solidify** modifier. These blue lines are the normal indicators and show which direction the face's normal points. Switching to **Wireframe** view (*Z*) and zooming in on the trouble area, it becomes apparent that the normals on the bad part are all pointing in. Normals should always point out.

On the left-hand side bar in the **Mesh Tools** section, there is a **Normals** section. There are two buttons there: **Recalculate** and **Flip Direction**

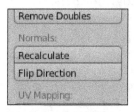

To practice selecting faces the **Flip Direction** button will be employed to selectively fix the normals:

1. On the bottom menu of the 3D View window, click on the Face select button:

2. If not in the **Wireframe** view, toggle it on (Z).

3. **Border Select** (B) the faces, whose normals are facing inwards:

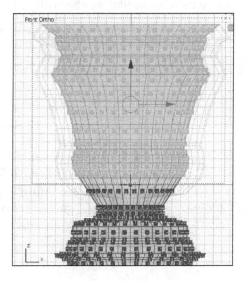

 Since face select is on any face whose center point is within the border select box will be selected. It is okay to over select slightly as long as the center points of the good faces are not within the selection area.

4. Click on the **Flip Direction** button in the left side bar of the 3D View window.

5. In the bottom menu bar of the 3D View window, click on the Vertex select button.

6. Exit the **Edit Mode** (*Tab*).

Now, **Solidify** works as expected and the vase has a 2 mm wall with the traced outline on the outside.

Making a solid base

Before continuing, now is a good time for another gentle reminder about saves. Use saves (*Ctrl + S*) and incremental saves (*F2, +*) from time to time.

A proper vase will hold water, so this vase will have to be modified so the base instead of having a 2 mm wall will be solid, and the bottom of the body and foot closed so it can hold water and print. This involves selecting the internal walls of the vase, which can be difficult to do without selecting the external walls as well, so a new selection operation will the used; **Select More**:

1. Apply the **Solidify** modifier.

2. Toggle to the **Wireframe** view on (*Z*).

3. Enter the **Edit Mode** (*Tab*).

4. **(De)select All** (*A*).

5. In the front view, zoom in on the base and identify a line of vertices, where the noses meet which are on the inside and carefully **Border Select** (*B*) the whole line. Try to identify a line that is not so close to external lines in this view to make the selection easy:

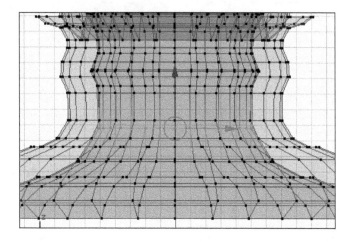

6. From the menu at the bottom of the View Panel, click on **Select More** or press *Ctrl* and the + key on the number pad:

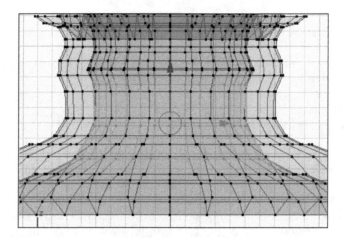

7. Keep expanding the selection (*Ctrl* + *NumPad* +) until the line of vertices on the very bottom of the vase is selected:

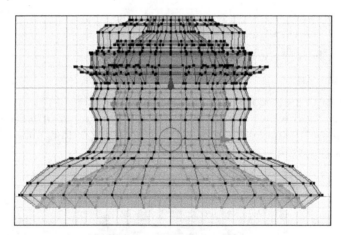

8. Begin the **Border Select** operation (*B*).

9. Click and hold the middle-mouse button and draw a box around the bottom few selected layers to deselect those points.

Using the *middle-click* to deselect, works with the **Circle Select** operation (*C*) as well.

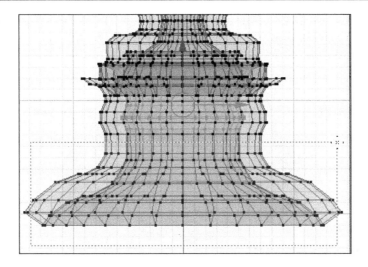

10. Continue to select more (*Ctrl + Numpad +*) and box deselect (*B, middle-click*) the points on the bottom or top of the selection area to avoid over selecting until all the vertices inside the base are selected, not including the vertices at the foot of the vase or the bottom of the inside of the body.

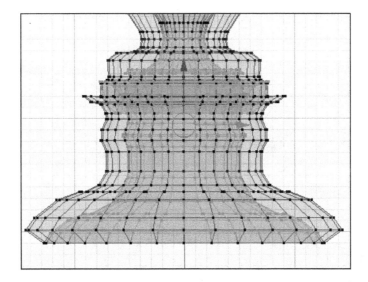

11. **Delete** the selected vertices (*X*).

The **Select More** option has an opposite in the **Select Less** operator, which intuitively is accessed from the same menu or by pressing *Ctrl* and the *NumPad* - key on the keyboard.

Switch to **Solid View** mode (*Z*) and rotate the view with the keys on the number pad to inspect the bottom of the vase. The shape has a big hole in it that will need to be closed to make this object printable:

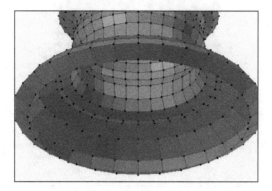

There is a trick for selecting whole loops like this very quickly, which will make filling in the hole easy:

1. With the view adjusted so that the hole in the bottom is visible, press *Alt* and *right-click* on an edge of the hole to select the whole loop.

2. Press *F* to create a new face with the selected vertices:

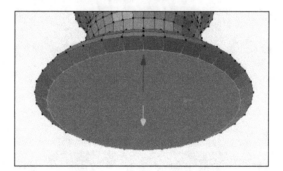

There is another hole inside the body that can be filled in the same way, but will need to be done in **Wireframe** mode (*Z*). Once in **Wireframe** mode, zoom in where the noses meet and look for the loop of points inside the body. Then, same as for the other area, loop select (*Alt* + *right-click*) the hole and make a face from the selection (*F*).

Whenever vertices, lines, or faces are deleted, there will likely be holes in the shape. It is important to catch and close these holes or there is a chance the shape will not print. Exploring how to find these problems in files downloaded online will be explored in a later chapter.

Printing the vase

The vase is now complete and can be exported. Exit the **Edit Mode** and with the final vase selected navigate to **File | Export | Stl (.stl)** in the menu bar. Then process in the usual way for printing. The example model had some overhang in the mouth area but printed okay:

Extra credit

There are often many techniques that can be employed to make the same thing. In fact, there is another way that this sort of vase could have been created using mostly just the techniques learned in the previous chapter. Starting with a cylinder at the bottom of the reference image, extrude the top slightly, and scale each extruded segment:

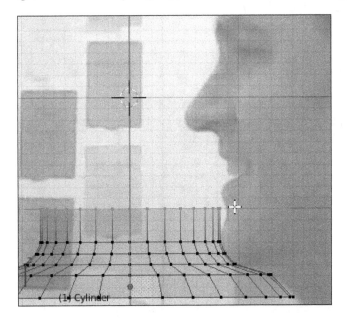

The disadvantage of this method is that it is difficult to get the same level of detail, and the wall inside the body may not be exactly 2 mm the whole way. It is left to the modeler to verify tolerances. The advantage of this method is that a clever modeler could take a picture of two different people facing each other and make a vase that looks different on each side. It requires some clever manipulation of the reference image to line up the features and could make an excellent gift.

Summary

Creating this vase taught many important 3D modeling skills and principles as follows:

- Importing a reference image
- The **Screw** modifier
- The **Solidify** modifier
- Normals, viewing, and fixing
- Deleting vertices
- Creating new faces from selections

The next chapter is another cool project that involves making something to match the dimensions of a real-life object, a useful, if somewhat silly, ring that can hold an SD card on a finger.

4
SD Card Holder Ring

Some 3D printers can print directly from an SD card instead of being hooked up to a computer all the time. But what if the computer used to prepare files is not close to the 3D printer? Some sort of SD card holder can be used in order to keep the card safe, but not to tie your hands up (in case you don't have any pockets available). A ring that you can put an SD card in is just the thing.

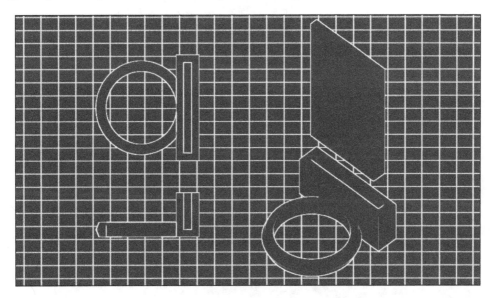

In this chapter an object will be modeled to real-life specifications. The challenges of matching real-life measurements with plastic shrinkage will be explored as well. In the end, a functional real-life useful object will be created, even though some would claim Blender isn't suited for CAD-like precision work.

As precise placement of new objects is important in this project, remember that if the 3D cursor is ever moved with an accidental left-mouse click, navigate to **View | Align View | Center Cursor** and **View All** or press *Shift + C* to put the 3D cursor back at the origin.

Taking measurements

This would be a very simple build using the tools taught in previous chapters. So this set of instructions will focus on using a different paradigm of designing. In this build, the **Apply** button in the modifiers will not be used so that the object remains modifiable until the end. This technique works well for simple shapes but can be a problem when the geometry of the model becomes increasingly complex.

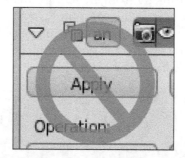

Before opening Blender, some measurements must be gathered. Careful measurements need to be taken of the finger that the ring will be on and of the SD card. A digital caliper is an excellent tool to take measurements with:

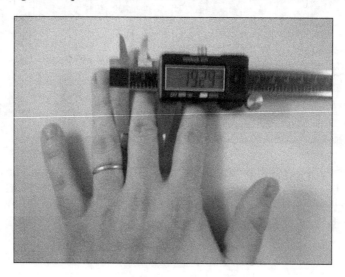

My middle finger measures 19.3mm at the widest point. If a digital caliper is not available, another way to measure a finger is to wrap a piece of paper around the finger, mark where it overlaps, and use a ruler to find your finger's circumference. Then consult the next chart to find out the standard ring size and diameter:

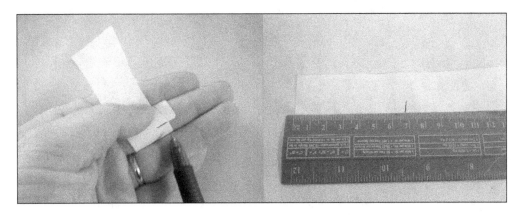

Circumference (mm)	Circumference (in)	Diameter (mm)	Radius (mm)	US Standard Ring Size
1.94	49.3	15.7	7.85	5
2.04	51.9	16.51	8.25	6
2.14	54.4	17.32	8.66	7
2.24	57	18.14	9.07	8
2.34	59.5	18.95	9.47	9
2.44	62.1	19.76	9.88	10
2.54	64.6	20.57	10.28	11
2.65	67.2	21.39	10.69	12
2.75	69.7	22.2	11.1	13
2.85	72.3	23.01	11.50	14
2.95	74.8	23.83	11.91	15
3.05	77.4	24.64	12.32	16

If all else fails, there is always trial and error. Make several test rings of various sizes, print them out, then see which one fits best. An adult male generally has a size of around 10; an adult female around a size 8, a child is around a size 5. If the test rings are made in the standard sizes and are marked, they could be kept for the future to model rings for friends and family.

Next, a standard SD card needs to be measured. Fortunately SD cards are all the same: 2.2 mm x 24 mm x 32 mm:

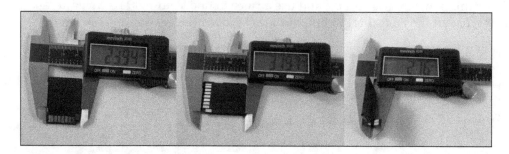

Modeling the finger

Now that all the measurements have been taken, it is time to go to Blender. Start a new scene; **(De)select All** (*A*) objects and **Delete** (*X*) them. Then save the scene (*Ctrl + S*) in a new directory under the `Makerbot Blueprints` directory called `Ch4 SD Card Ring` and name the project file as `SD Card Ring`.

1. **Add** (*Shift + A*) a cylinder. In the side bar to the left, change the options for the cylinder. Change the value of **Vertices** to 64 so the cylinder is smoother than default. Change the value of **Radius** to half the diameter of the measured finger. Change the value of **Depth** to 10:

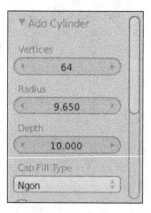

 Blender can take simple equations in these boxes, so instead of dividing the diameter in half to get the radius, simply type the diameter followed by /2 and it will divide it by 2 for you. So, in the preceding example, typing 19.3/2 resulted in the value of **Radius** being shown.

2. Select the **Object** tab (orange cube) from the properties on the side bar to the right, rename this cylinder to Finger as this cylinder represents the finger in the build space:

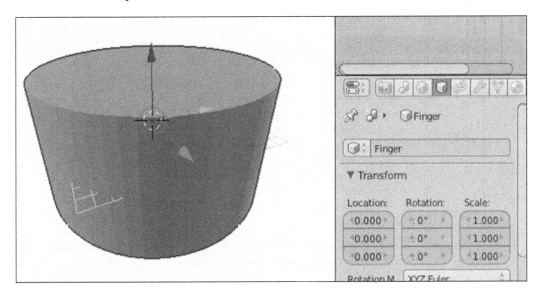

3. **Add** (*Shift + A*) another cylinder. Blender remembers the settings that the last object was created with, and if another object is created of the same type, those settings will be applied to it. So the new cylinder is of the same radius, depth, and number of vertices as the Finger cylinder and is therefore in the same exact place until the settings are changed. Increase the value of the **Radius** by 2 of the **Finger** cylinder and change the value of **Depth** to 4:

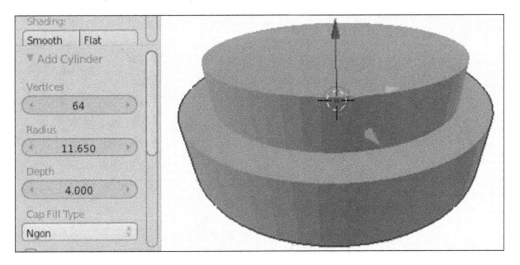

Again, adding a value of 2 to the existing diameter is as simple as clicking on it and appending +2 to the end and then pressing *Enter*.

4. In the **Object** tab rename this cylinder to Ring.

 The XY plane is the logical floor of the build space that passes through the origin. Blender places a (small) XY plane in the view space to help visualization. It is, of course, completely imaginary and objects can be built through this floor. In fact, by default, new objects are placed somewhat under the XY plane. However, respecting the XY plane can provide a solid base of reference for building objects if needs be, especially when multiple objects will be placed with precision.

5. Jump to the **Front** (*Numpad 1*) **orthographic** (*Numpad 5*) view. Notice that the ring is sitting half above and half below the origin.

Putting the ring on the floor

The goal of this section will be to place the important objects so they sit on the XY plane without going below it.

1. With only the ring object selected, begin the **Grab/Move** (G) operation.
2. Press the Z key to lock the ring's movement along the z axis.
3. Type 2 to move the ring up half its depth.
4. Press *Enter* to end the grab operation.

 Like scaling, you can lock movement to a single axis. This works regardless of the view and ensures that the movement is predictable.

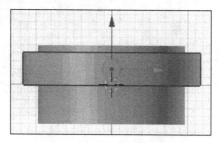

This puts the `Finger` cylinder in an odd place, but as it is still both above and below the ring cylinder, it is not important and can just be left where it is.

Making a test ring

1. Similar to what has been done in other projects, add a Boolean modifier to the ring and difference the `Finger` object from it. The `Finger` object obscures the effect of this operation, so hide it. One way to do this is to select (*right-click*) the `Finger` object and navigate to **Show/Hide | Hide Selected** from the menu at the bottom of the **3D View Panel Object** or press the *H* key. Another way is to click on the icon that looks like an eye in the **Outliner** on the line for the `Finger` object:

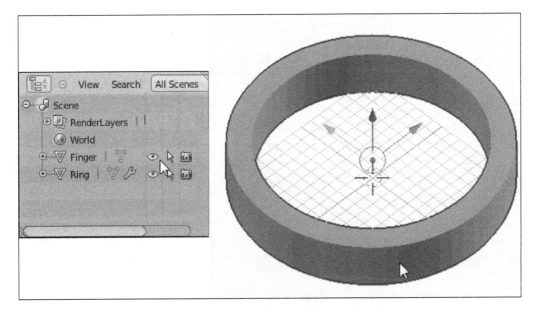

2. The ring is now perfectly serviceable, but there's no harm in making it a little more appealing. Enter the **Edit Mode** (*Tab*). Again, notice that without applying the modifier, its effects are undone while in **Edit** mode. Loop Cut (*Ctrl + R*) the cylinder around the *middle-click* and *left-click* twice when selecting, so the slice does not slide up or down:

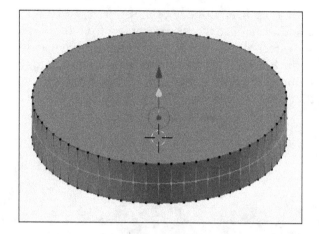

3. **Scale** (*S*) the newly sliced ring slightly outwards. The scaled ring of points should not extend more than 1 mm (one small grid square in the background) larger than the top and bottom disks of the ring shape:

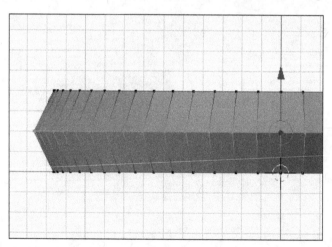

Scaling is an operation for which precise measured adjustments like this are very difficult to accomplish. This is one of the places where Blender falls short as a CAD tool. Just something to keep in mind when designing.

4. Exit the **Edit Mode** (*Tab*). This is a good time to save (*Ctrl + S*) the changes. Export an STL of the ring by navigating to the **File | Export | Stl (.stl)** menu options. Name the exported file `Test Ring.stl`. Then print the test ring and try it on:

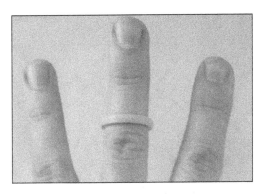

It might come as a surprise that the ring probably doesn't fit. If printing in ABS, the plastic shrinks, as that is one of the properties of ABS. Or there may be a print setting such as filament diameter that isn't correctly set. Also, internal rings are known to be a little small. Whatever the reason, this ring size isn't going to work; so it's a good thing that it was tested while the ring was still small.

Resizing the test ring

Now we are back to Blender to make a new test ring. This time, instead of starting from scratch, reload the project and resize the existing ring. Care has to be taken when resizing the ring to insure that it doesn't extend below the XY plane.

1. Unhide the `Finger` cylinder by clicking on the eye icon in the **Outliner** on the line for the `Finger` object.
2. **Select All** (*A*) objects.
3. Begin the **Scale** (*S*) operation.
4. Press *Shift + Z* to lock the scale operation to all but the z axis.
5. Type `1.03` to scale the selected object just slightly larger.
6. Press *Enter* to end the scale operation.

Pressing *Shift* and *X*, *Y*, or *Z* key when in an operation such as scale or grab, locks the operation to everything but the selected axis, regardless of view. With rotation, however, the effects are less predictable.

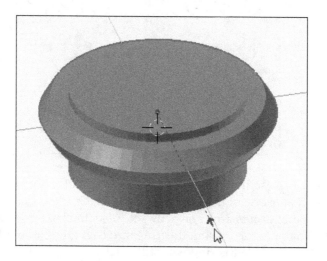

 The logic behind using 1.03 as a scaling factor is that past experimentation has proven this to be the right amount to overcome ABS shrinkage. It also coincidently brings the new ring very close to the next ring size. For completeness, it will be mentioned that scaling by 0.97 will approximately undo a scale of 1.03, if making the object smaller is desired. If a more exact measurement is desired, the exact amount of scale can be calculated by measuring the ring after printing with the calipers, calculating a ratio of the differences, and solving for the corrected scale factor. This is an exercise left for the reader.

7. **Save** (*Ctrl* + *S*) and export the model again. Print it out and test again. If it is still too small, repeat the process of scaling the ring up until it can be worn and removed comfortably. If during the process of printing, the printing software displays a message to the effect that the ring is not touching the bottom of the build platform, it is an indication that at some point locking the scale operation to all but the z axis was skipped and the ring now is falling below the XY plane. It is not a good idea to let this happen. It would be preferable to go back to an earlier iterative save and use a larger ring size to start with.

When a comfortable ring has been achieved, continue with the project.

Adding an SD card holder

1. **(De)select All** objects (*A*) and press *Shift + H*. Remember that *Shift + H* hides all but the selected object; however, as no objects are selected, all objects will be hidden to clear the way for what is to come. **Add** (*Shift + A*) a cube to the scene.

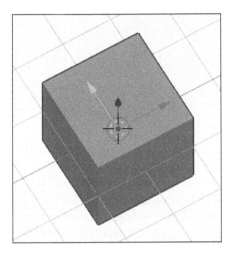

It is interesting to note that the default cube created by blender is not a unit, that is to say, its sides are not all 1 mm. They are in fact 2 mm each. This makes the default cube less desirable for precise scaling. Fortunately, it is an easy thing to make this cube a unit cube; however, first its dimensions can be used to move it on to the XY plane easily. Only this time a trick will be employed to control how the cube scales in the future. Instead of moving the whole cube up, the individual points will be moved up in the **Edit Mode** so that the object's origin is left where it is.

 Object origin was a problem in the last project causing the object's axes to be skewed in relation to the world orientation. In that project, the rotation was corrected to fix it. In this project, the origin will be intentionally skewed to produce a desirable result.

2. Enter the **Edit Mode** (*Tab*).
3. Select all of the (*A*) points.
4. Begin the **Grab/Move** (*G*) operation.
5. Press the *Z* key to lock the movement in the z axis.
6. Type 1 to move the selected points up 1 unit.

7. Press *Enter* to end the grab operation.

8. Exit the **Edit Mode** (*Tab*).

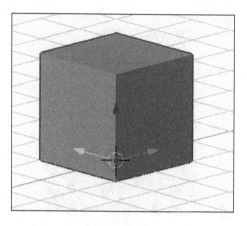

The cube now sits nicely on the XY plane, but more than that, its origin is also on the XY plane on the bottom face of the cube. Experimentally **Scale** (*S*) the cube and move the mouse. Notice that with the cube's origin where it is the scale operation is only scaling the cube above the XY plane. Undo (*Ctrl + Z*) any experimental scaling before continuing.

It is now time to make the cube unit a unit cube. This consists of simply scaling (*S*) the cube by 0.5. Now all the sides of the cube are exactly 1 unit long. A unit cube is very useful for making precise objects such as a virtual model of an SD card.

1. Begin the **Scale** (*S*) and press *Enter* to end it without changing anything.

2. In the **Resize** options in the side bar to the left, edit the dimensions of the scale operation based on the measurements of the SD card taken earlier.

3. Change the value of **X** to 2.2, the value of **Y** to 24, and the value of **Z** to 32.

4. Select the **Object** tab (orange cube) from the properties on the side bar to the right and rename this cube to SD Card.

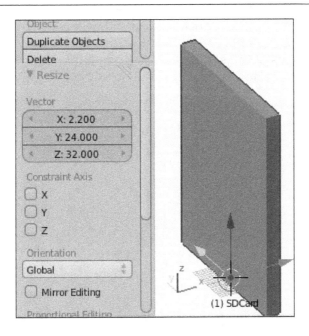

(1) SDCard

5. **Add** (*Shift* + *A*) a new cube. Repeat the procedure of entering the **Edit Mode** (*Tab*) and moving (*G*) its vertices 1 unit along the z axis (*Z*), then exit the **Edit Mode** (*Tab*). **Scale** (*S*) the cube by .5 to make it a unit cube. Then use the **Scale** (*S*) operation again and type in the dimensions **X**: 6.2, **Y**: 28, **Z**: 12. Rename this cube to SD Holder:

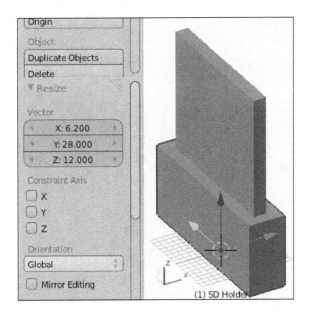

(1) SD Holder

6. This makes the holder 2 mm thicker than the SD card all around and short enough to easily remove the SD card while tall enough to hold it securely. Add a Boolean modifier to the SD holder and difference the SD card from it. **Hide** (*H*) the SD card and rotate the view to check the SD holder.

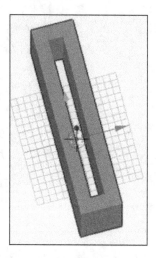

7. A hole all the way through was not the desired effect, as the card might fall out of the ring. Since the Boolean operation has not been applied yet changes can still be made. Unhide the SD card by clicking on its eye icon in the **Outline** panel. Select the **SD Card** object and move (*G*) it 2 units along the z axis (*Z*). Hide the **SD Card** object again and check the SD holder:

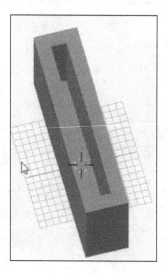

This looks much better.

Adding the SD holder to the ring

Unhide the ring by clicking its eye icon in the **Outline** panel. Select **SD Holder** and **Grab/Move** (G) it in the x axis (X) until it is at the front of the ring. Depending on the view when the last operation was done, a problem may have been observed:

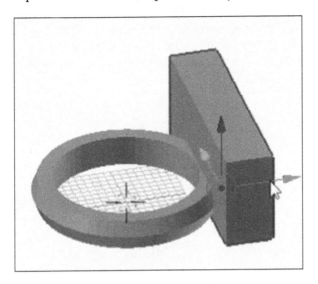

What happened to the hole for the SD card? Because the Boolean modifier was not applied, the hole stayed right where it was where the hidden **SD Card** object resides:

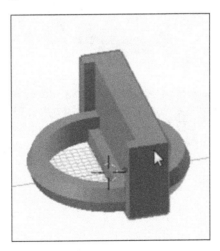

As amusing as this can be, this is an undesired behavior. There are two options. Either the Boolean operation can be applied before moving, or both objects can be moved together. As the stated goal of this project was to not use the **Apply** button, the choice is clear. Undo (*Ctrl + Z*) the move operators until the hole is where it belongs, in **SD Holder**. Then unhide the **SD Card** object by clicking on its eye icon in the **Outliner** view. Select both **SD Holder** and **SD Card** by right-clicking on one, then holding *Shift* and right-clicking on the other. Then **Grab/Move** (*G*) and move them both in the x axis (*X*) until they are at the front of the ring with the ring intersecting **SD Holder**. Make sure there is a good connection here without interfering with the finger hole.

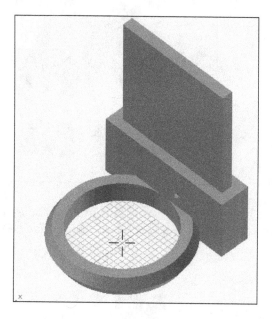

Hide the **SD Card** object now that everything is in place. Select the ring and add a Boolean modifier to union the ring and **SD Holder** together. Once the Boolean modifier is added, **Hide** (*H*) **SD Holder** to make inspecting the new object easier. Rotating the view, a blemish will be discovered. Inside the holder for the SD card, the ring is invading the hole for the SD card. If it isn't clear, switch to the **Wireframe** view (*Z*) and look at the ring in **Front** (*Numpad 1*) orthographic view.

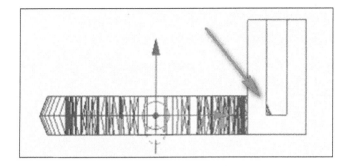

The solution is to change around the various Boolean operations that are being done. Right now, the **SD Card** object is being differenced from **SD Holder**; the resultant shape is being united with the ring. The correct order is the **SD Holder** being united with the ring, and then the **SD Card** being differenced from the resultant shape. Because these are modifiers that haven't been applied, they can be changed while the objects are still hidden using only the **Outline** and **Properties** panel, both of which are in the side bar to the right.

1. In the **Outline** view click on **SD Holder**.
2. With **SD Holder** selected, click on the **X** button on its Boolean modifier to remove it.

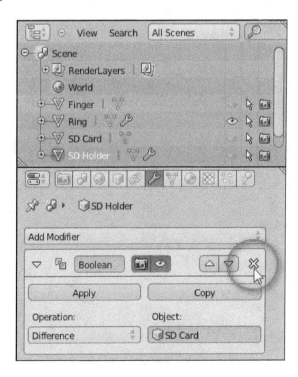

3. In the outline view click on **Ring**.

4. Add a Boolean modifier and **Difference** the **SD Card**.

The **Ring** now has three Boolean modifiers on it, all at the same time creating the desired object.

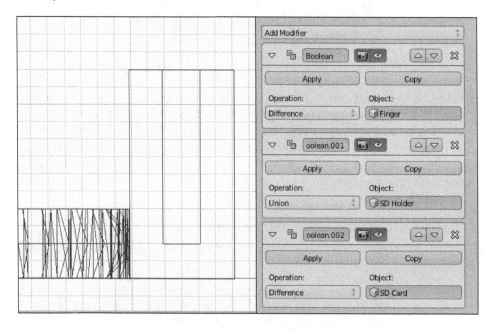

Through careful planning, the bottom of the model is flat and ready to go. All that is left now is to export (**File | Export | Stl (.stl)**), print out, and use the new SD Card holder ring.

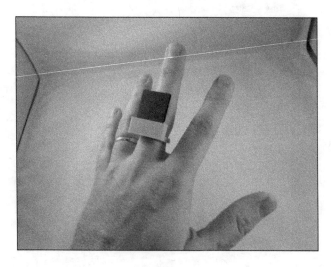

Extra credit

Leaving modifiers unapplied becomes undesirable when the object's geometry gets too complex. Simply performing Boolean operations on cubes together may not cause a crash, but adding Boolean operations to multiresolutioned objects and moving them around or changing the order of their modifiers has been known to slow down or even crash even the heartiest of systems.

Try to see how quickly Blender will slow down by making several cubes, adding multiresolution modifiers to increase their polygon count and Boolean them together all without pressing the **Apply** button. Add more and more multiresolution cubes and move them around to see when performance becomes choppy. It may be surprising how little time it takes.

With the measurements of an SD card available, any object can be turned into an SD card holder. Why not make an SD holder key chain or an SD holder that clips to a breast pocket? Add an SD holder to a mini mug or vase to turn decorative items into something functional. Measure pencils and pens and create a desk organizer as well. Locking the SD card may also be needed in some applications—an exercise left to the reader.

Summary

Leaving modifiers in place is a technique that allows for a high degree of flexibility that is particularly desirable for something that stands a good chance of being customized. If someone else wanted their own SD card holder ring, it would be trivial to resize the ring and adjust to position of the SD Holder.

Also in this chapter, Blender's tools for precise placement were employed to make an object match real-life measurements. Blender may never be able to compete with other CAD programs for precision, nevertheless with some careful planning and clever manipulation, Blender is capable of extremely precise modeling.

The next chapter is a huge challenge where the tools and techniques learned so far will be applied to making a modular multi-part toy with moving parts. The modeling will be simple, but the result will be impressive as the ways that parts can be joined together and designing those joints to be printable will be explored.

5

Modular Robot Toy

3D printers can make a number of useful and practical things. But what's the point if you can't have fun too? This blueprint is for a 3D printed poseable robot in multiple parts that connects together with 3D printed connectors. It will be connected with two types of connectors, a pin connector that will do the majority of the joints, and a ball and socket connector for the head.

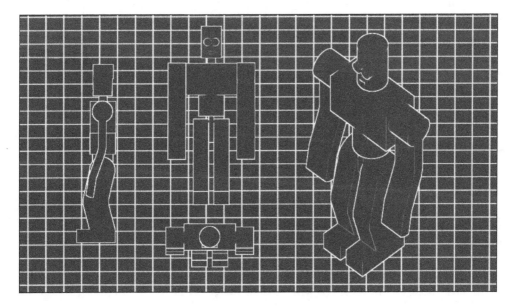

Before this project begins a lot of planning has to be done. Often projects list this start long before the modeling program opens up. This should be considered normal for any design project. Fortunately in this case the planning is already done and the modeling can begin.

The straight pin connector used here is loosely based on a connector created by *Tony Buser* on *Thingiverse*; modified to fit this project. This connector relies on the flexible nature of the plastic to get the head into a smaller hole, until it is past the opening, and can spring back to hold the part in place. When designing this sort of connectors it is generally a good idea to lay them down so that the arms flex along the printed layers and not across them, which might cause the layers to separate. Hence, instead of trying to incorporate the pegs into the body, they'll be printed separately and added afterwards. This increases the bulk of the robot because there needs to be holes on both sides of the connections, but it does improve the reliability of the print.

This blueprint will rely heavily on exact object placement. This is also the largest project in the book, so remember to save frequently, and do incremental saves at every new header section.

Making the connector

Start a new blender project, clear the workspace, and save the file. Create a new directory under the `Makerbot Blueprints` directory in **Documents** called `Ch 5 Robot Toy`. Enter the directory and save this file as `Peg.blend`.

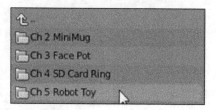

 Instead of designing everything in one file in this project, multiple files will be generated. This is a good way to keep parts organized, and keep the outliner from being overrun with objects. This is especially a good idea when the parts being designed might be reusable in future projects.

1. Add (*Shift* + *A*) a **Cylinder** to the scene. Change the cylinder's **Radius** to 3.5 to bring it to a total diameter of 7mm.

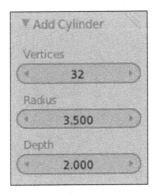

2. In the **Properties** pane under the **Object** tab change the name of the cylinder to **PegBody**.

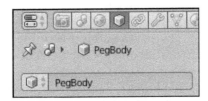

3. **Rotate** (*R*) the **PegBody** around the y axis (*Y*) 90 degrees.

 If the rotation is accomplished with the key sequence *R*, *Y* then type 90 and press *Enter* to end the rotation action, it doesn't matter what view it is done in, the result will always be the same.

4. In the **Front** (*Numpad 1*) **Orthographic** (*Numpad 5*) view, enter **Edit Mode** (*Tab*). With all points selected grab and move (*G*) all the points in the x axis (*X*) 1 unit by typing the number 1.

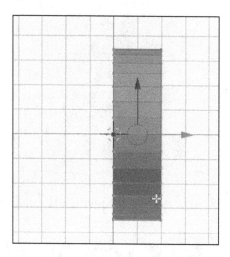

5. Exit **Edit Mode** (*Tab*). Then in the **Object Modifiers** tab in the **Properties** pane press the **Add Modifier** and choose the **Mirror** modifier. The mirror modifier makes a mirror copy of the object through its own origin. Same is the case with the SD card holder, moving the points in edit mode kept the object origin, and moved all the points to one side of that origin so that the mirror reflection won't overlap.

6. Looking at the default options for this modifier it looks like the mirroring should be happen along the x axis, which is exactly what is desired, but there is no visible effect to the **PegBody**. What is happening? This is the same problem that the screw modifier had before. The object has been rotated so its local axis is out of line with the global axis. The procedure for fixing this is the same as well. Apply (*Ctrl + A*) **Rotation and Scale** to normalize the object's axis again, and the mirror modifier will work as expected.

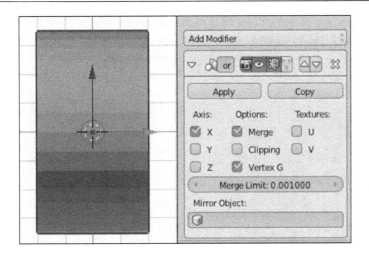

7. Enter **Edit Mode** (*Tab*) and then switch to **Wireframe** mode (*Z*). Deselect all points (*A*). **Border Select** (*B*) all the points on the right side of the **PegBody**. Then **Grab/Move** (*G*) and in the x axis (*X*) enter 5 units.

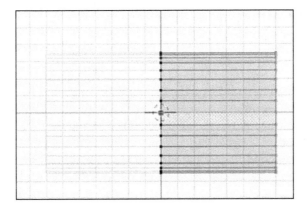

8. The next few steps will be to build a geometry that will be modified to match the plan. First **Extrude** (*E*) a 0.75 length, then **Extrude** (*E*) another 0.75 length, and finally **Extrude** (*E*) a 1.5 unit length.

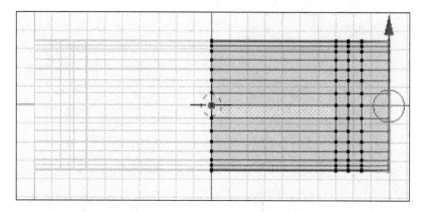

Blender will let you type any number in when doing an operation. However there are a few caveats that may confuse the new user. First of all there is no obvious prompt echoing what is being typed. The *Backspace* key will reset the typing portion of the action so the user can retype his parameter, but it will not undo the movement until something else is typed in.

Fortunately there are a few tricks that will help. First of all along the bottom of the 3D view during any operation the default menu is replaced with some information about the current operation being performed. Next, if things get really unmanageable pressing the *right-click* or *Esc* will cancel the operation. However, if the *Enter* key or *left-click* is clicked on completing the operation, there is always undo (*Ctrl + Z*).

9. With the right-most end of the **PegBody** still selected **Scale** (*S*) it down until it is only about 6 units tall. Remember with the scale operation the exact measurements are difficult, so use the mouse and estimate. Changing the view and zooming in on the area being worked on can help.

10. Deselect all point (*A*) and **Border Select** (*B*) the points on the second and third column of point from the right. **Scale** (*S*) them in all but x axis (*Shift* + *X*) until they are about 8 units across total.

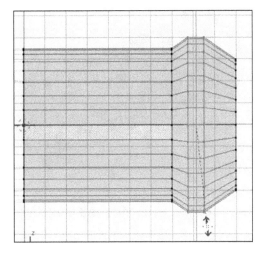

There is one last thing to do. Because this shape is being mirrored, the leftmost face of the shape is not necessary however; it will not be removed by the mirror modifier. Leaving it in won't do much harm, it will simply cause the peg to have an internal wall that won't be seen, but may interfere with Boolean operations later so it's best to get rid of it.

11. **Border Select** (*B*) the points on the right side of the body, in the middle where the mirror happens, press **Delete** (*X*) and select **Faces** from the menu.

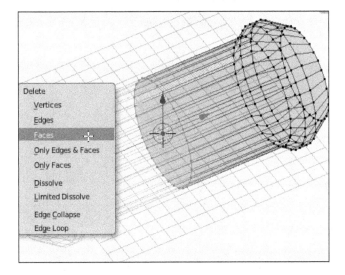

12. Exit **Edit Mode** (*Tab*) and the general shape of the PegBody is complete.

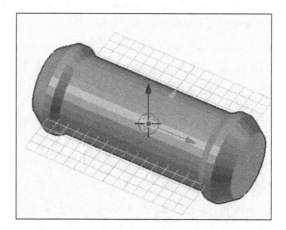

13. Remember to **Save** (*Ctrl + S*) frequently and to do incremental saves once in a while.

Splitting the connector

Now for splitting the connector perform the following steps:

1. Add (*Shift + A*) another **Cylinder**. Change the cylinder's radius to 1.5 (so it is 3 units across). In the **Properties** panel under the **Object** tab rename the **Cylinder** to PegSplit.

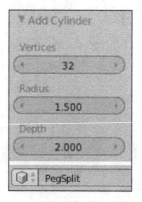

2. **Rotate** (*R*) the PegSplit around the y axis (*Y*) 90 degrees. Enter **Edit Mode** (*Tab*) and with all points selected **Grab/Move** (*G*) and move the points along the x axis (*X*) 3 units.

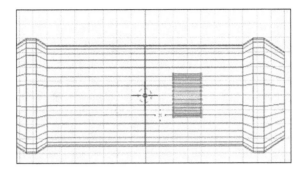

3. Deselect all points (*A*) and **Border Select** (*B*) the points on the right side of the PegSplit. **Grab/Move** (*G*) and move them 7 units along the x axis (*X*).

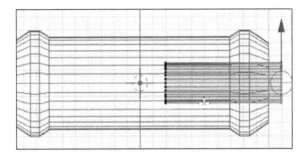

4. Jump to the right view (*Numpad 3*). Clear the selection (*A*) and use the **Border Select** (*B*) to select the points at the top of the PegSplit that are within 1 unit on either side of the z axis line. See following illustration:

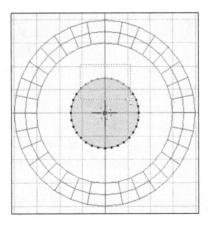

5. **Extrude** (*E*) these points to 4 units. Do the same on the bottom of the `PegSplit`. The result will be a shape that sticks out of the top and bottom of one side of the `PegBody`.

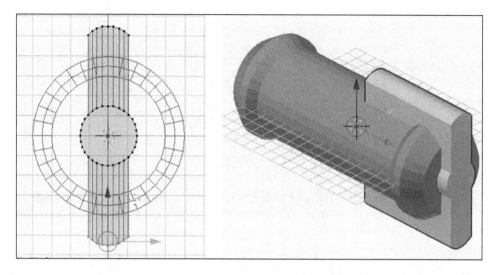

6. **Hide** (*H*) the **PegSlice**. Select (*right-click*) the `PegBody` and in the **Properties** panel, **Modifiers** tab, add a **Boolean** modifier to the `PegBody`. Change the **Operation** to **Difference** and select the **PegSlice** from the menu.

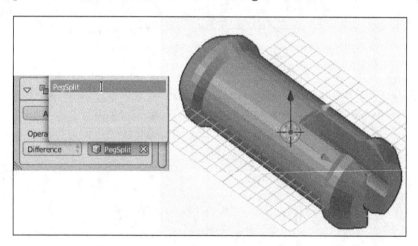

The mirror modifier is still active, its effect hasn't been made permanent with the **Apply** button yet, so why is the split only appearing in one side of the peg? The reason is because the order of the modifiers matter. Here, the main shape is being mirrored first, and then Boolean is applied. Actually it should be the other way around. It is very easy to change the order of modify operations.

In the **Modifier** options press the button with a triangle pointed upwards on the Boolean modifier to move that modifier up the stack. Then the Boolean will be applied first, and the peg will be split on both ends after the mirror.

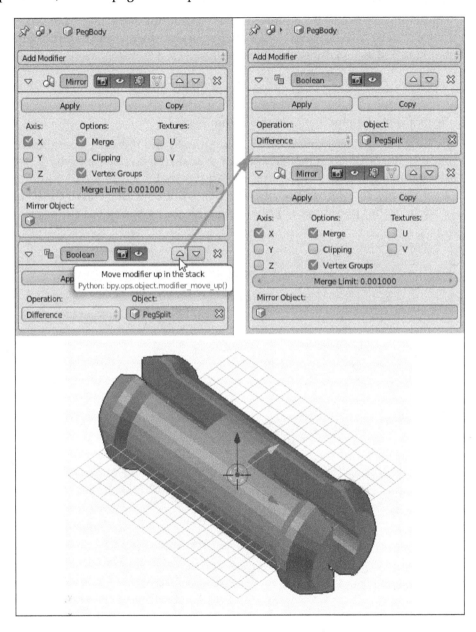

Building a printable peg

Theoretically this is a perfectly functional peg. However it lacks two things that are important for 3D printing; a flat surface and little overhang. This can be easily fixed by flattening the top to be symmetrical at the bottom of the peg. Some of its roundness will be lost, however that's not important since it will still fulfill its function, and it will also be hidden inside the objects it is holding together, so it will never be seen.

1. Create (*Shift* + *A*) a **Cube**. Rename the cube to Peg.

2. **Scale** (*S*) the peg by 0.5 to make it unit.

3. **Scale** (*S*) again and immediately press *Enter* to end the scale operation, and modify the options on the left side bar to scale the box by 22 in the **X**, 10 in the **Y** and 6 in the **Z**. This will produce a box that is slightly wider and longer than the peg but slightly shorter. The peg will be sticking out of the top and bottom of the cube shape slightly.

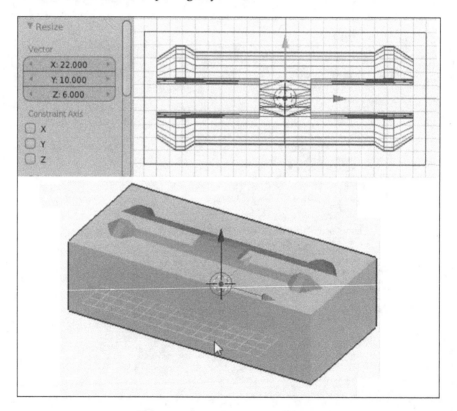

4. Hide the `PegBody` either by selecting (*right-click*), and using the hot key to **Hide** (*H*) it or clicking the eye icon on the `PegBody` line in the **Outliner** view.

5. Select (*right-click*) the cube shape and add a **Boolean** modifier to it. Change the operation to **Intersect** and select the **PegBody**.

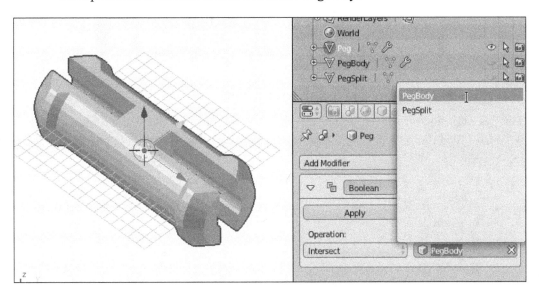

There are only three types of Boolean operations with intersection being the last one to be introduced. Boolean operations take two objects and combine them in different ways. Union joins both of the shapes into one shape, which was used on the Mini Mug and SD card holder ring. Difference, used earlier in this blueprint and in the SD card holder, as well takes the second object and subtracts its shape from the first one. Intersection produces a shape from where the two shapes overlap. In this case the resulting shape was the shape of the PegBody minus the top and bottom where the cube didn't go.

With Difference it is important, which object is being modified and which object is being used as a parameter. If your difference object A from object B the results will be very different than if you difference B from A. However, with union and intersection, the order of the shapes doesn't matter. You can add a Boolean modifier to either shape with the other one as a perimeter, and the result will be exactly the same. In this case the cube was intersected with the PegBody because the PegBody already had several modifiers on it and the cube would have had no purpose but to be a parameter otherwise.

6. With the peg complete this is the right time to export it. In the **File** menu in the **InfoBar** panel choose **File | Export | STL**. Make sure the **Ch 5 Robot Toy** directory is the active one and name the file Peg.Stl.

Putting a hole in our pocket

The peg is done and printable, but the shape of the hole for the peg will need to be defined. This hole will be Boolean differenced from other shapes later to make a hole the perfect size and shape for the peg.

1. With the peg visible add (*Shift + A*) a **Cylinder**.

2. Change its **Radius** to 3.7 so that there is 0.2mm of clearance around the peg in the hole.

3. Rename this cylinder to PegHole.

4. **Rotate** (*R*) it 90 degrees around the y axis (*Y*).

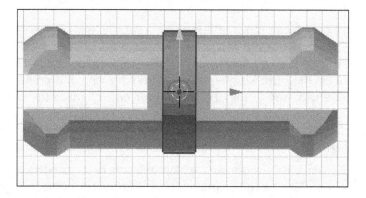

 It is a good idea to make your holes just slightly bigger than the object that is fitting into them. 0.2mm tends to be a good tolerance to design for.

Shaping the PegHole is going to be very similar to making the PegBody.

1. Apply (*Ctrl + A*) **Rotation** and **Scale**.

2. Jump to the top view (*Numpad 7*).

3. Enter **Edit Mode** (*Tab*).

4. With all (*A*) points selected **Grab/Move** (*G*) all the points in the x axis (*X*) 1.

5. Without exiting edit mode add the **Mirror** modifier.

Adding modifiers while in the edit mode has always been possible, but many modifiers have no visible effect until edit mode is exited. However mirror is not one of those modifiers.

1. Deselect all points (*A*).

2. **Border Select** (*B*) all the points on the left side of the PegHole.

3. **Delete** (*X*) the **Faces**.

4. **Border Select** (*B*) the point on the right.

5. **Grab/Move** (*G*), and move them along the x axis (*X*) 5 units.

6. Then, extrude (*E*) 0.75, extrude (*E*) 0.75 length, and extrude (*E*) 2.

    The breaks will link up with the peg and the hole will extend just past the end of the peg.

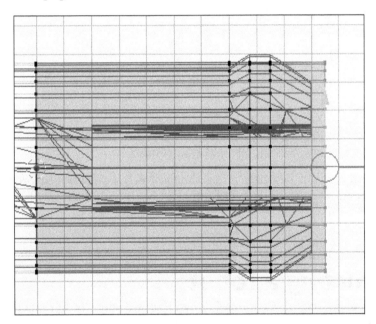

7. **Scale** (*S*) the end to about 6 units across.

8. Deselect (*A*) and border select the second and third rows of points and **Scale** (*S*) them in all but the x axis (*Shift + X*) until they are approximately 0.2mm wider than the bulges in the peg. The slopes of the shapes should run parallel.

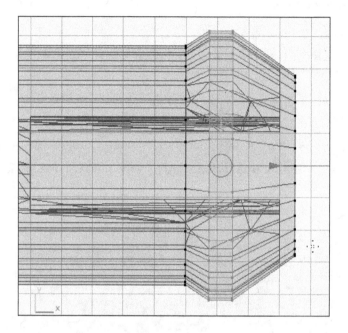

9. When the shape is complete exit **Edit Mode** (*Tab*).

If at this point the peg or peg hole is less than perfect, and since this project is so expansive, a version of the **Ch5 - peg.blend** can be downloaded from Thingiverse at http://www.thingiverse.com/thing:90754 so that the rest of the project can be completed without needing to backtrack.

Constructing a robot

The peg is done and its hole is defined. Now, to build something that will use them. This new build will need the PegHole from the other project.

1. Start a new project (*Ctrl + N*). Clear the scene and **Save** (*Ctrl + S*) this project in the same directory as Peg.blend but name it Robot1.blend.

2. In the Info panel on the top of the screen select **File | Append** or press *Shift + F1*.

3. Select the `Peg.blend` file. Select the **Object** directory and select the
    PegHole. The PegHole from the previous project will be brought into
    the Robot1 project.

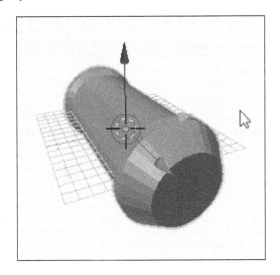

The robot as planned will need four pegs. Two at the shoulders, one at the torso, and one at the hips. To make things easier the one at the hips will be this peg at origin so it doesn't have to move and the robot will be built around it. Rename this `PegHole` to `PegHip`.

Engineering the body

1. **Hide** (*H*) the `PegHip` for now.

2. Jump to the front (*Numpad 1*) orthographic (*Numpad 5*) view.

3. Add (*Shift + A*) a cube.

4. Scale (*S*) it `0.5`. **Scale** (*S*) again and press *Enter* to end the scale operation, then type in the following dimensions in the resize operation options: **X:** `40`, **Y:** `20`, **Z:** `24`.

5. Name this cube `Body`.

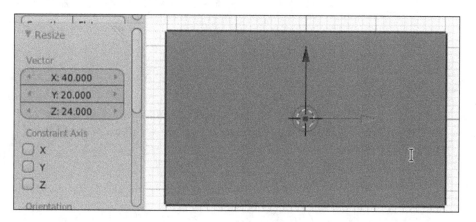

6. **Grab/Move** (*G*) the body along the z axis (*Z*) `40` units. (Remember to type `40` in to get the units exact.)

Because of the neck joint's location in the print it can be made with a ball joint without causing an overhang problem.

1. Begin by adding (*Shift + A*) a **UV Sphere**.

2. Change the sphere's **Size** to `4` (which is the radius, so the sphere will actually be 8 units across).

3. Rename the sphere to `Neck`.

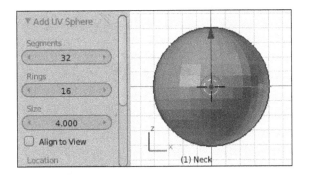

4. **Grab/Move** (*G*) the neck along the z axis (*Z*) `64` units.

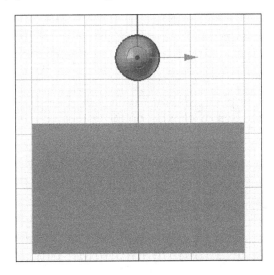

5. In **Edit Mode** (*Tab*), **Wireframe** view (*Z*), **Border Select** (*B*) all the points on the bottom four rings of the sphere.

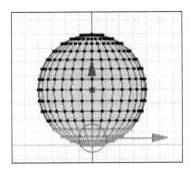

6. **Extrude** (*E*) these points 10 units. **Scale** (*S*) the extrusion by about 1.2 to give the neck a stable base.

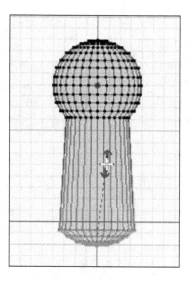

7. With the neck complete exit **Edit Mode** (*Tab*). In the outline view find the **PegHip** and unhide it by clicking on the eye icon on its line.

8. Select (*right-click*) the **PegHip**.

After starting a duplicate action it behaves like a grab operation on the new object. As long as the duplicate action is not completed with the *Enter* key or left-mouse button the new object can be moved in the same way as with a grab.

9. Duplicate it (*Shift + D*) and move it along the z axis (*Z*) 28 units.

10. **Rotate** (*R*) it 90 degrees around the y axis (*Y*).

11. Rename this new object `PegTorso`.

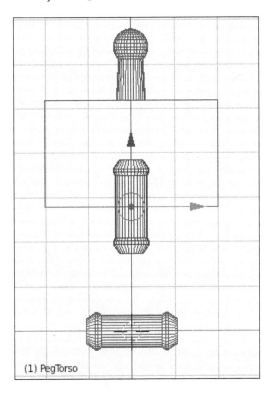

(1) PegTorso

12. Select (*left-click*) the **PegHip** again.

13. **Duplicate** (*Shift + D*) it again and move it along the z axis (*Z*) 12 units.

14. Complete the action with the *Enter* or left mouse button.

15. **Rotate** (*R*) this peg 90 degrees around the y axis (*Y*). This peg will be made into two and placed in the left and right shoulder.

16. **Duplicate** (*Shift + D*) the peg and move the new duplicate 20 units along the x axis (*X*).

17. Select (*right-click*) the original peg that is still in the middle of the body, and **Grab/Move** (*G*) it (not duplicate this time) -20 units along the x axis (*X*).

18. Name the left peg `PegLeft` and the right peg `PegRight`.

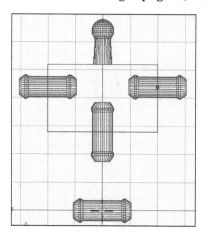

19. Select (*right-click*) the **Body** and begin to add **Boolean** modifiers to it.
20. First add a **Boolean** to **Union** the **Neck** to the body.
21. Then add **Booleans** to **Difference** the **PegLeft**, **PegRight**, and **PegTorso** from the body.
22. Finally, hide everything but the body (*Shift + H*) and admire the completed body shape.

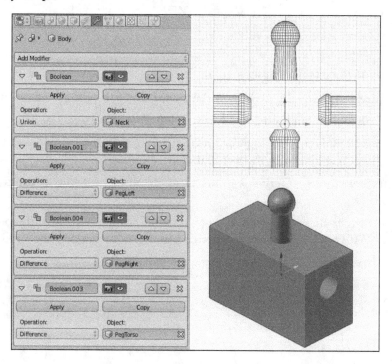

Creating the torso

1. **Add** (*Shift* + *A*) a cylinder. Change its dimensions to **Radius**: 10, **Depth**: 16. Rename this cylinder Torso.

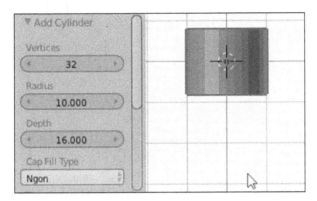

2. **Grab/Move** (*G*) it *20* units along the z axis (*Z*).

3. **Add** (*Shift* + *A*) another cylinder, this time with dimensions **Radius**: 8, **Depth**: 4. **Rotate** (*R*) it 90 degrees around the y axis (*Y*). Rename this to Hips.

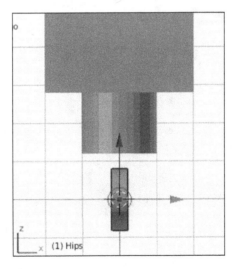

4. Enter **Edit Mode** (*Tab*) and in the front view (*Numpad 1*) in **Wireframe** mode (*Z*), **Box Select** (*B*) all the points on the top half of the hips including the points at the midpoint.

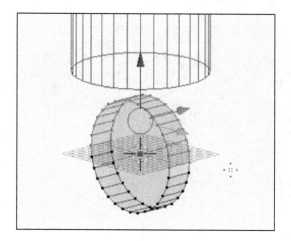

5. **Extrude** (*E*) these points 14 units. They should go into the torso.

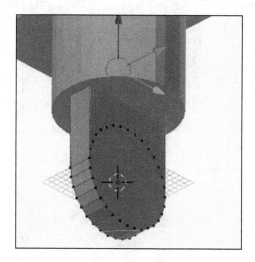

6. Exit **Edit Mode** (*Tab*).
7. **Hide** (*H*) the hips.
8. Select (*right-click*) the torso, and add a **Boolean** modifier to **Union** the torso to the **Hips**.
9. Then add another **Boolean** modifier to **Difference** the **PegHip**.
10. Lastly, add one more **Boolean** modifier to **Difference** the **PegTorso**.

The torso is complete:

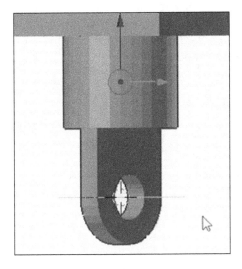

Making an arm

Because of the symmetry of the arms and legs it will be possible to model only one of each, and use the mirror modifier to make the other. As a result, only one arm will be modeled at this time.

1. **Add** (*Shift* + *A*) a cylinder, edit its parameters to **Radius:** 8, **Depth:** 14.

2. Rotate (*R*) it 90 degrees around the y axis (*Y*).

3. Rename the cylinder Arm.

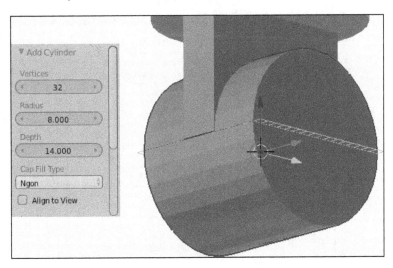

The move operation can be used with a technique similar to the way the scale operation has been used before; start the operation, end it without changing anything, then edit the parameters. Since most of the movements before this were done in only one axis, there wasn't much advantage to this technique, but since the arm has to be moved in the y and z at the same time, placement will be accomplished easier using this technique:

- Begin the **Grab/Move** (*G*) operation.
- Press *Enter* or the *left-click* to end the operation without moving the mouse.
- Edit the values in the left sidebar to **X:** 27, **Y:** 0, **Z:** 44.

The arm will be properly placed in one quick move.

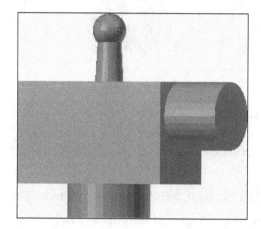

1. Enter Edit Mode (*Tab*) and switch to **Wireframe** view (*Z*).
2. From the right side view (*Numpad 3*) select only the vertices at the bottom of the arm being sure to select at least 2 units width of points.

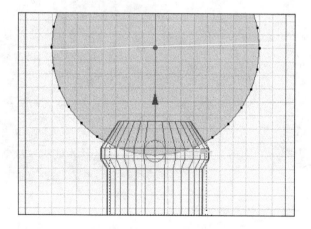

3. **Extrude** (*E*) these points 20 units.

4. Again, **Extrude** (*E*) another 4 units.

5. **Rotate** (*R*) the selected points 20 degrees.

6. **Grab/Move** (*G*) them -1 unit along the y axis (*Y*).

7. **Extrude** (*E*) the points another 20 units and then **Extrude** (*E*) another 15 units.

8. **Move** (*G*) the points along the y axis (*Y*) 3 units.

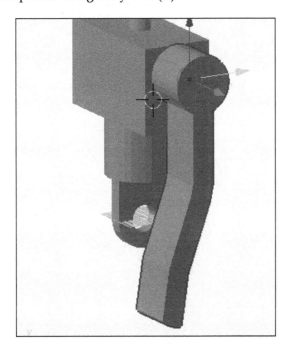

9. Lastly add a **Boolean** modifier to **Difference** the appropriate shoulder peg (**PegLeft** or **PegRight**) from the arm. Use **Wireframe** mode (*Z*) to check that the Boolean operation worked as expected.

Shaping the leg

Now to model a simple leg:

1. Start by adding (*Shift* + *A*) a cylinder. **Radius:** 10, **Depth:** 12.

2. **Rotate** (*R*) it 90 degrees around the y axis (*Y*) and name the cylinder leg.

3. **Move** (*G*) the leg along the x axis (*X*) 8 units.

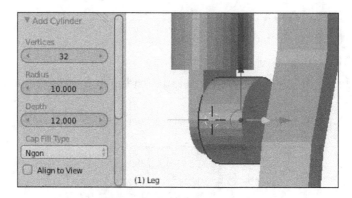

4. In **Edit Mode** (*Tab*), **Wireframe** view (*Z*), right side view (*Numpad 3*) select all the points on the bottom half of the leg.

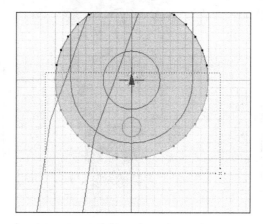

5. **Extrude** (*E*) 15 units.
6. **Scale** (*S*) the selection along the z axis (*Z*) by 0 units to flatten the bottom.

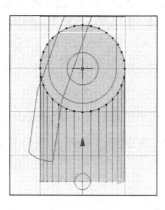

7. **Extrude** (*E*) the selection by 4 units. **Extrude** (*E*) again 20 units, then **Extrude** (*E*) again by 10 units.

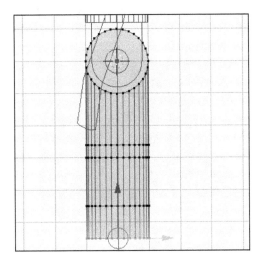

8. Select only the points on the knee (see the illustration below) and **Grab/Move** (*G*) them along the y axis (*Y*) by -5 units.

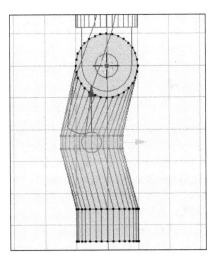

9. From the side view it is not apparent how flat all the points on the side of the leg are, but adjusting the view it becomes clear.

10. Select the points on the front of the foot and **Extrude** (*E*) those by 10 units. It might be easier to select these points by switching back to solid view mode (*Z*) for a moment and selecting (*right-click*) them one at a time while holding the *Shift* key (remembering to clear the selection (*A*) before starting).

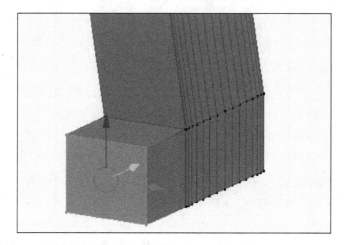

11. Exit **Edit Mode** (*Tab*).

12. Add a **Boolean** modifier to difference the **PegHip** from the leg.

13. Check in **Wireframe** mode (*Z*) that the Boolean modifier was correctly applied.

Forming the head

When the head is made it will need a hole in the back of it so the neck can be pushed in but that will hold the joint snugly. Unlike the peg holes a tighter fit is preferable for ball joints. To avoid stressing the part there must also be a way to guide the ball joint in that starts wide and narrows before the ball. Because of the importance of planning the hole properly the hole will be the first thing made.

1. For the neck hole **Add** (*Shift* + *A*) a cube.

2. Rename it Neckhole and **Grab/Move** (*G*) it 64 units along the z axis (*Z*).

3. **Scale** (*S*) the cube by 3.

Since this cube was not unitized (scaled by 0.5) firstly, this brings its size to sized 6 cube in the middle of the neck.

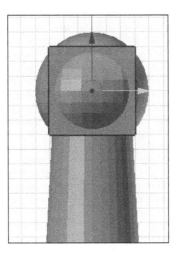

The next couple of moves will shape the cube around the neck hole. The shape will be irregular so it will be accomplished with a series of discrete moves.

1. In the right side view (*Numpad 3*) in **Edit Mode** (*Tab*), use **Wireframe** view (*Z*) to select all the bottom points.
2. **Grab/Move** (*G*) them `-10` units down the z axis (*Z*).
3. Select the back points and move (*G*) them `6` units along the y axis (*Y*).
4. Select the front points and move (*G*) them along the y axis (*Y*) `-2`.
5. Finally select the top points and move those `2` units in along the z axis (*Z*).

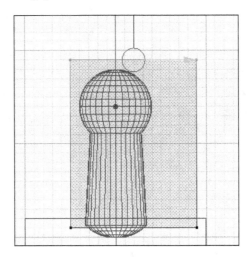

6. While still in edit mode and wireframe view jump to the front view (*Numpad 1*).

7. Select all the bottom points.

8. **Scale** (*S*) them along the x axis (*X*) about `1.2` so that the neck is mostly inside the box with only a little bit of the sphere up top sticking out on both sides.

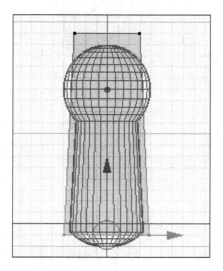

9. Jump to the top view (*Numpad 7*).

10. Select all the points on the back and **Scale** (*S*) them along the x axis (*X*) `1.2`.

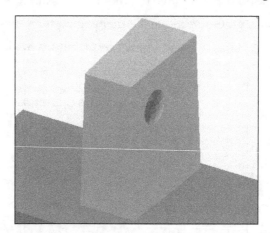

11. Exit **Edit Mode** (*Tab*).

12. Add a Boolean modifier to **Union** the **NeckHole** to the **Neck**.

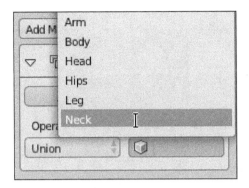

13. Now that the **NeckHole** is complete **Hide** (*H*) it for later.

14. **Add** (*Shift + A*) a cylinder and change its dimensions to **Radius:** 8, **Depth:** 24. Name it Head.

15. Start and end a move (*G*) operation and edit the parameters to **X:** 0, **Y:** -2, **Z:** 68.

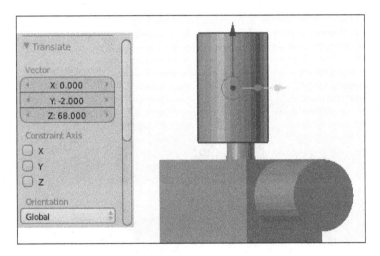

16. **Add** (*Shift + A*) a cylinder that has radius 3 and depth 8.

17. Name it Eye.

18. **Rotate** (*R*) it *90* degrees around the x axis (*X*).

19. Using similar steps to the movement of the head start a move operation (*G*), end it, and edit the parameters to **X:** 3, **Y:** -6, **Z:** 68.

20. Duplicate (*Shift* + *D*) the eye and move the duplicate along the x axis (*X*) -6 units.

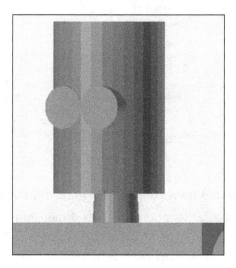

21. Finally **Add** (*Shift* + *A*) a cube.

22. Name it Mouth.

23. Start and end a move operation and edit the parameters to **X:** 0, **Y:** -10, **Z:** 60.

24. **Scale** (*S*) it by 0.5. Start and end another **Scale** (*S*) operation and edit the parameters to **X:** 4, **Y:** 2, **Z:** 2.

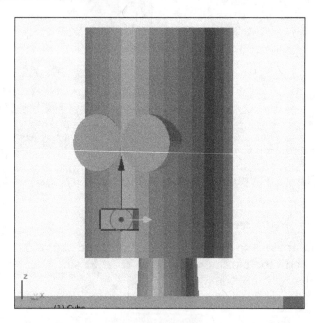

25. **Hide** (*H*) the **LeftEye, RightEye,** and **Mouth.**

26. Select (*right-click*) the head and begin to add modifiers to the head.

27. **Boolean Union** the **Eyes. Boolean Difference** the **Mouth** and **NeckHole.**

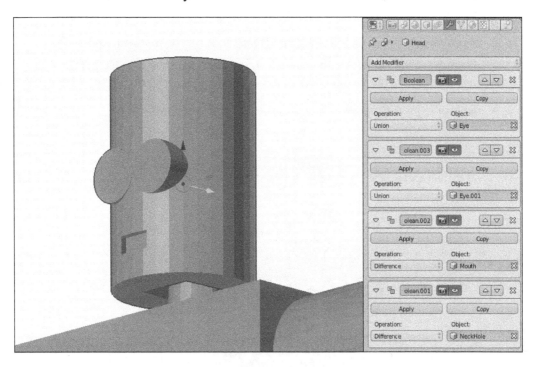

Assembling the parts to print

A gentle reminder to **Save** (*Ctrl + S*) and perform incremental saves is in order. Hopefully, frequently saving is a habit at this point.

Building the parts like this is good for getting proportions and alignment correct, but is not appropriate for printing. Each part will be prepared for printing individually using a new view mode called **Local,** accessed from the **View | View Global/Local** menu options at the bottom of the 3D View panel or by using the slash on the number pad (*Numpad /*).

1. Select the head and **Duplicate** (*Shift + D*) it.

2. Name the duplicate `PrintHead`.

3. Enter local view (*Numpad /*).

The only object visible will be the **PrintHead** that was the last object selected. This is different than hiding all other objects. This can be confusing since any attempts to modify other objects won't work as expected until local view is exited. Always look for the label **(Local)** on the view description in the upper-left of the 3D View panel.

4. Apply all the modifiers to the **PrintHead**, so it can be manipulated alone.

    Because of the neck hole the current orientation isn't the best for printing. Fortunately it was designed with a flat printable surface on the top of the head.

5. **Rotate** (*R*) the **PrintHead** 180 degrees around the y axis (*Y*).

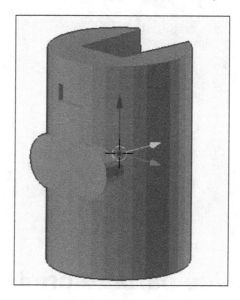

6. Choose **File | Export | STL** and save this part in the **Ch 5 Robot Toy** directory as **Head.STL**.

7. Hide (*H*) the **PrintHead** and exit local view (*Numpad /*).

8. Select (*right-click*) the body.

9. Same as with the head duplicate (*Shift + D*) it and rename the duplicate PrintBody.

10. Enter local view (*Numpad /*) with the PrintBody.

11. **Apply** all modifiers.

12. Fortunately this piece is printable as it is. In fact any change in orientation would make it less printable. So simply **File | Export | STL** this part as **Body.STL**.

13. **Hide** (*H*) the **PrintBody** and exit local view (*Numpad /*).

14. Select (*right-click*) the torso.

15. Repeat the steps of duplication (*Shift + D*), renaming **PrintTorso**, applying all modifiers, and entering local mode (*Numpad /*) to analyze it.

16. The torso clearly won't print the way it is, but rotating (*R*) 180 degrees around the y axis (*Y*) will fix it.

17. Do the manipulation, and **File | Export | STL** the **Torso.STL**.

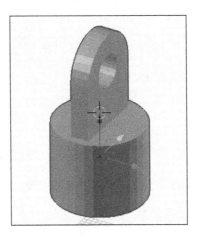

18. Then **Hide** (*H*) the **PrintTorso** and exit local view (*Numpad /*).

> If it is not clear why the naming convention of putting "Print" first is being used find the `PrintHead`, `PrintBody` and `PrintTorso` in the Outliner view. The Outliner view organizes things alphabetically so giving these parts all the same beginning makes them easier to find in the Outliner. It is a simple little trick to keep things of like type organized though it is somewhat backwards of English conventions.

Next is the arm:

1. **Duplicate** (*Shift + D*), name it `PrintArm`.
2. Apply all the modifiers and bring into local view (*Numpad /*).
3. Rotate (*R*) -90 degrees around the y axis (*Y*).
4. Before this part can be exported it needs to be mirrored for the other arm. Add a **Mirror** modifier.

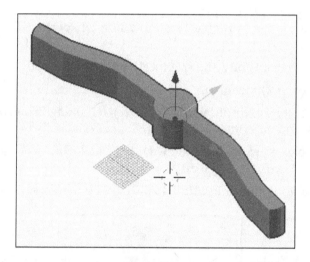

After doing the mirror operation instead of having two arms there will be a double sided arm joined at the shoulder. To fix this:

1. Enter **Edit Mode** (*Tab*).
2. Select all points (*A*) and move them in the x and y (*Shift + Z*), until there are two arms.
3. However the arms are rather spread out, and will be difficult to print on most printers. So **Rotate** (*R*) all the points 90 degrees around the z axis (*Z*) as well to make the result a little more compact.

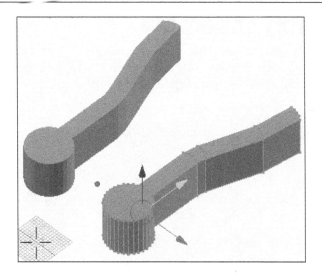

4. Exit **Edit Mode** (*Tab*) and **File** | **Export** | **STL** this as Arms.Stl.

5. Then hide (*H*) **PrintArm** and exit local view (*Numpad /*).

Finally, the leg in a similar fashion to the arm:

1. Duplicate (*Shift + D*), rename Printleg, and view the new leg locally (*Numpad /*).

2. Apply modifiers.

3. Oriented for print by rotating (*R*) -90 degrees around the y axis (*Y*).

4. Finally, the mirror modifier is added, and edited (*Tab*) to make the separate parts.

Potentially the legs could be printed standing up since they have a flat bottom. It would be printable but tall, and the layers would line up differently. In the end whether the legs are laid down or stood up can be a matter of preference for the reader.

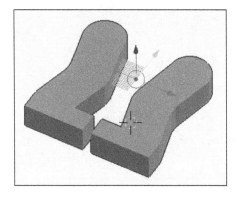

Printing and assembly

Print Head.STL, Body.STL, Torso.STL, Arms.STL, Legs.STL, and four copies of Peg.STL. Snap three of the pegs into the holes in the body. Snap the torso onto the peg at the bottom of the body. Push the last peg through the hole in the bottom of the torso.

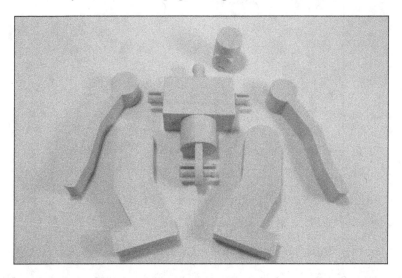

Snap the legs into the hip peg on either side. Snap the arms. Press the head into the neck, join firmly being careful not to break the neck joint. If it does break remove it with a pair of needle nosed pliers and glue it back in place with superglue. Allow it to dry and then try again. Then pose and play with the new robot.

Extra credit

Making a robot toy was a huge undertaking. But now that it's done why not go off the pattern and make a new style that is compatible with the old? There's a reason this project was named Robot1.blend. Your own robot doesn't need to follow the strict alignment that this one did, that was just for tutorial purposes. Make a rounder robot based on a sphere or one that's all cubes. Make a robot with spikes and blades for hands. Just be sure to plan a flat surface for printing and remember the overhang.

If multiple robots are made to use the same sized connectors, then the parts can be mixed and matched between them. With only four different robots trading heads, bodies, arms, and legs there are a potential 256 different robot combinations that could come from them. That's a lot of play for a little bit of modeling.

Summary

In this chapter, we learned a new way of connecting printed objects with custom connectors. Peg and ball joints were taught and their utility was demonstrated. The tolerances, or how much space needs to be in-between printed parts, was also explored.

In the next chapter another way of making parts interact using gears will be explored.

6

D6 Spinner

Imagine you're playing a board game, but can't find a six sided die. Printing one is no good, since printed dice haven't been proven to be balanced. A dice is not the only way to choose a number from one to six. With 3D printing a custom enclosed spinner can be printed, so you'll never be at a loss for a random number.

The toy robot in the last chapter isn't the only way a 3D printed object can be made to move. In this chapter we'll explore the creation of an object designed to move in the manner of a simple machine, with a simple gear powered by a common spring, as found in a pen or mechanical pencil.

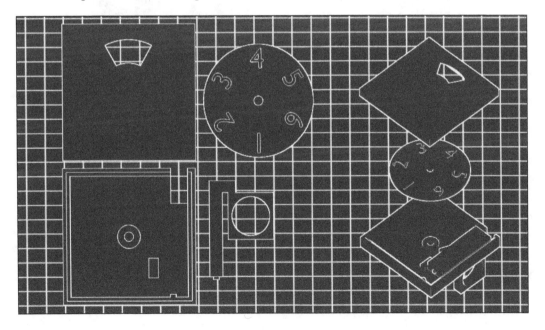

Blender does not, by default, have the tools that allow it to make gears with ease. But one of Blender's strengths is that it is modular, and can easily have its functionality extended. Some of these extensions are included in Blender's installation, and just waiting to be activated. One of them adds gears to the objects that can be created.

This project will start by extracting the spring from a mechanical pencil or pen. Then the spring will be measured, and a model will be built around it using an add-on for Blender that allows the creation of gear shapes.

Extracting the spring

Small springs suitable for jobs like this can be found inside spent mechanical pencils or ball point pens. Some mechanical pencils may need to be broken to get to the spring inside.

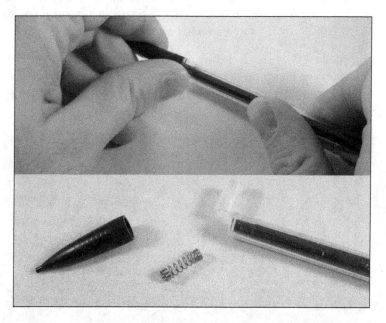

Once a spring is acquired it will need to be measured carefully. Measure the outside and inside diameter of the spring, as well as its length when fully extended and fully compressed. Most springs of these kinds will have similar measurements to the one in the illustration. If the spring used is different, then the dimensions in the project will have to be adjusted.

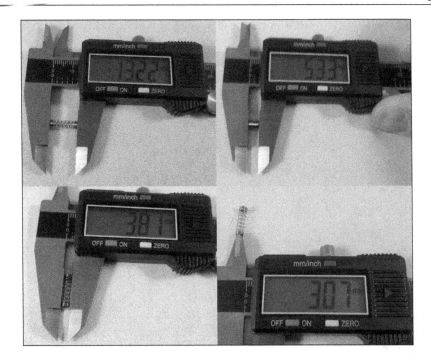

Starting the project

Start a new scene and clear it. **Save** (*Ctrl + S*) the scene in a new subdirectory of the **Makerbot Blueprints** directory named Ch 6 Dice Spinner. Name the new project Dicespinner.Blend. Remember while working to **Save** (*Ctrl + S*) frequently and do incremental saves occasionally.

Modeling the spring

Perform the following steps for modeling the spring:

1. **Add** (*Shift + A*) a cylinder and change its dimensions to the dimensions of the measured spring when extended (remembering the measured by typing the diameter followed by a /2, and blender will divide it by 2 and give the radius.) In the example the diameter was 3.8mm or 1.9mm for the radius and the depth was 13.2mm.

2. **Add** (*Shift + A*) another cylinder changing the depth to the length of the compressed spring, 5.3 in the example. Name these objects Springextended and Springcompressed. The internal measurement will be important later but unused in this step.

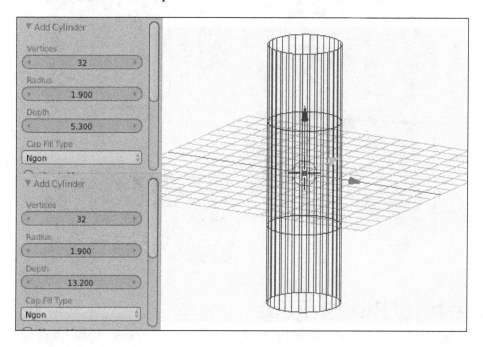

3. **Hide** (*H*) the spring shapes for later.

Defining the boundaries

This chapter will employ a new method for changing the dimensions and placement of objects that is perhaps simpler than methods in the previous chapters, but less natural. In the **Properties** side bar, usually hidden on the right side of the **3D View** panel, the location and dimensions of objects can be directly manipulated.

1. **Add** (*Shift + A*) a cube. Name this cube BoxExterior. Bring up the properties sidebar either by clicking on the tab with a plus on the right of the **3D View** panel or by pressing *N*. Locate the **Dimensions** under the **Transform** section. Change the **Dimensions** to **X:** 70, **Y:** 70, **Z:** 16.

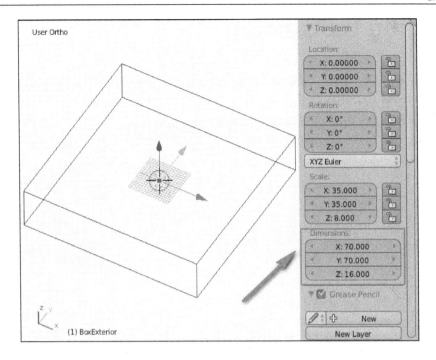

2. **Add** (*Shift* + *A*) another cube. Name it BoxInterior. In the **Properties** sidebar (*N*), change the dimensions to **X:** 62, **Y:** 62, **Z:** 13.

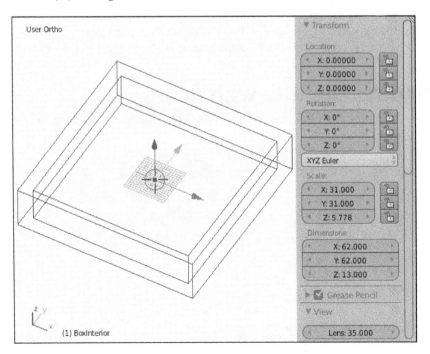

Building the spinner

Let's build the spinner:

1. **Add** (*Shift* + *A*) a cylinder. Name the object Spinner. In the **Properties** sidebar (*N*), change the dimensions to **X:** 58, **Y:** 58, **Z:** 2.

2. In the properties the location can also be altered so change the Z-location to 3 to move the spinner up a bit.

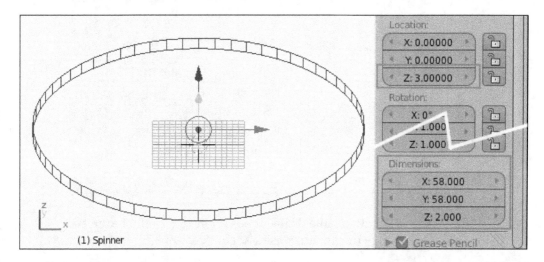

Now is the time to add the gear. There is a plug-in that adds gear primates packaged with Blender, but it is not activated by default so we will need to take care of it.

Extending Blender with gears

1. In the menu bar navigate to **File | User Preferences**. Click on the **Addons** tab.

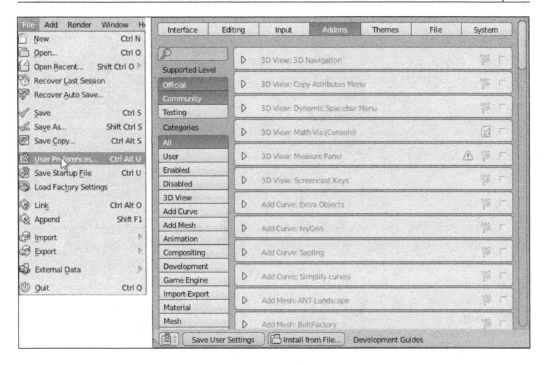

2. On the left-hand side click on the **Add Mesh** button to filter the add-ons.

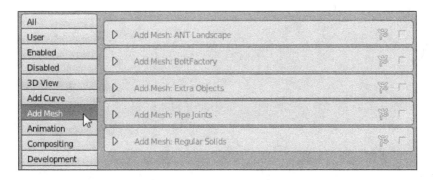

3. On the right find the **Add Mesh: Extra Objects** and click on the check box to enable it.

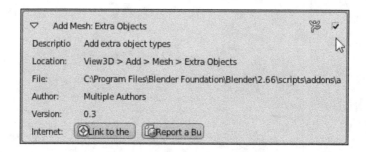

4. Then click on the **Save Settings** button to make this change default so the next time Blender is opened it will be remembered and close the **User Settings** window.

5. Now under the **Add Object** (*Shift + A*) **Mesh** menu there will be a new submenu for **Extra Objects** presenting many different new objects one of which is **Gears**. Add a **Gear** object now.

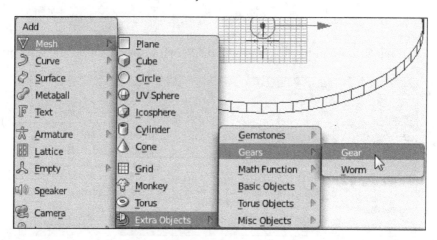

In the right sidebar, the newly added gear has many options, more than any other object used before. But as usual these must be set before any other action is performed or they'll be gone. The best thing to do at this point, as with any new tool, is to experiment with each setting. Change them to 0 and change them to 10. Make predictions and test them. Learn how this tool works.

If the gear seems somewhat inconsistent or incomplete that is one of the problems of working with an add-on, and the reason why add-ons like this are not enabled by default. The gear generated has a hole in the middle; it is not a closed object. To fix that the addendum setting can be set so that the points all converge in the middle. However, there is something about this that isn't right. Try to predict what the problem might be before it shows up later.

Another oddity of the gear add-on is that its width is not its total height, but rather is only half its height. This will just have to be accounted for.

Finally, it should be mentioned that this add-on does not make the ideal shape for a gear, known as an involute gear. This gear will work fine for this application, but is not necessarily suitable for more technical applications. Other modeling programs have better modules for doing all sorts of gears including involute gears. A search on Thingiverse will also reveal many different gear projects and is worth the exploration.

1. Change the settings of the gear so it is 5mm tall and has six 10mm long teeth with the following settings:

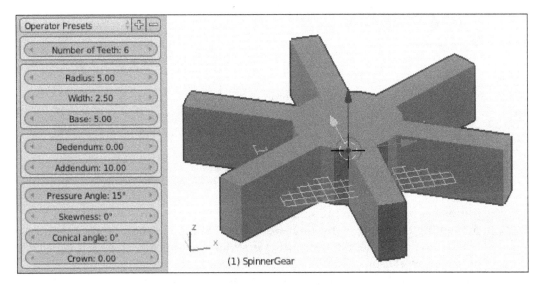

2. When the gear is the desired shape name it `SpinnerGear`. The `SpinnerGear` should overlap into the spinner above it slightly, which will make them Boolean together.

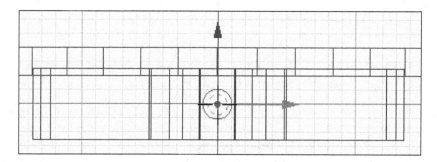

Adding a gear to the spinner

1. Select the spinner and add a **Boolean** modifier. Union the **SpinnerGear** to the spinner. At first it may look good, but hide all but the spinner (*Shift + H*), and apparently something is wrong. In fact the Boolean modifier gives a warning message:

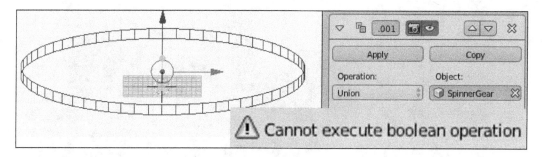

(Emphasis added.) The problem is that the gear has some bad geometry preventing the Boolean modifier from functioning. Particularly the junction where the points meet in the middle is still not technically closed. It is just a bunch of overlapping points. This is bad because technically there is still a hole in the mesh. Fortunately this is an easy one to fix.

2. Unhide the **SpinnerGear** and in **Edit Mode** (*Tab*) select all the points (*A*). Then in the left-side bar find and click on the button to **Remove Doubles**.

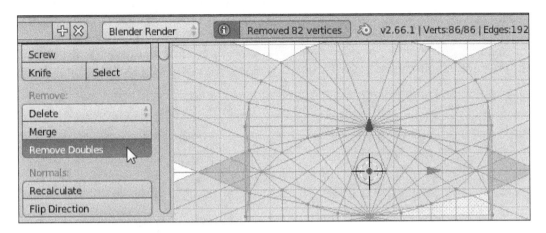

A prompt along the top should indicate that some amount of duplicate pixels were removed, and when returning to object mode the spinner's Boolean modifier will automatically update, the error message will go away, and the gear and spinner will be united into one piece. If the **SpinnerGear** is hidden (*H*), apparently it will mean that the Spinner now has six gear teeth attached to its base.

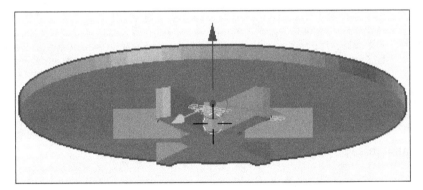

Spinning on a peg

In order for the spinner to spin, it will need a peg to spin on. The problem is that a traditional joint wheel and peg doesn't allow the object to lay flat for 3D printing no matter how it is turned. So making something like this work requires a little bit of creativity. Instead of a peg through the spinner a peg can exit out one side of the spinner and a hole can be made in the other side for another peg to mate into it.

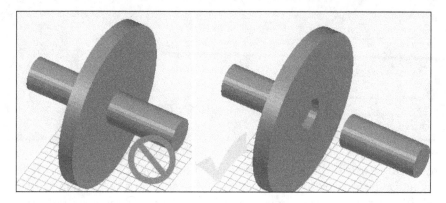

 The trick is that any time two objects are connected after printing, the goal is to think about friction. Since this spinner needs to spin the goal will be to build something that will hold the spinner in place enough that it will not shift position too much but loose enough to minimize friction so it can spin freely. Later on the box will be designed with extremely tight tolerances to increase friction since no movement is necessary.

1. **Add** (*Shift + A*) a **UV Sphere** and adjust it so it has a radius of 2 and is moved (*G*) -4 down the z axis (*Z*) so it is sitting somewhat below the main spinner. Name this sphere SpinnerPeg.

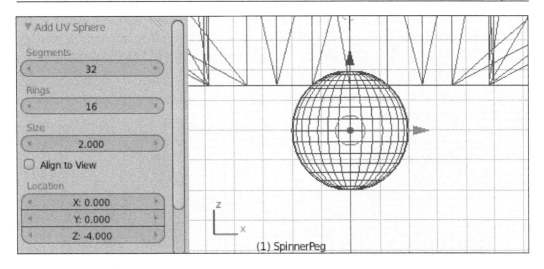

2. Enter **Edit Mode** (*Tab*), select the vertices on the top half of the sphere including the equator and **Extrude** (*E*) it 2.

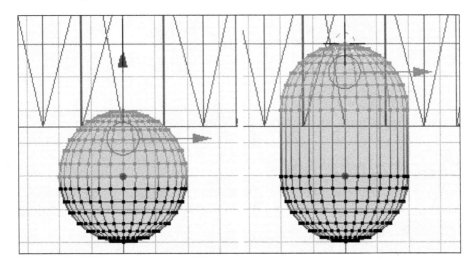

3. Then select the spinner and Boolean union the **SpinnerPeg** to the **Spinner**.

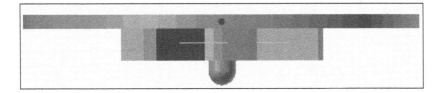

Now a hole needs to be made in the top of the spinner for a similar peg that will come from the top of the box to keep the spinner in place. As was mentioned earlier this hole will need to keep the peg in place but not stop it from spinning freely.

4. **Add** (*Shift + A*) a **Cylinder**. Set its radius at 2.4, its depth at 5, and move (*G*) it 4 along the z axis (*Z*).

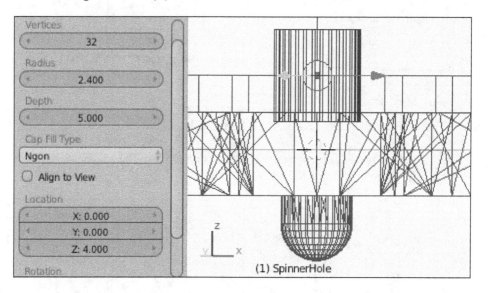

5. Select the **Spinner** and **Boolean Difference** the **SpinnerHole** from it.

Adding the numbers

The last thing to add are numbers to the surface of the spinner using Blender's **Text** object. The text object operates slightly different than other mesh objects. **Add** (**Shift + A**) a **Text** object.

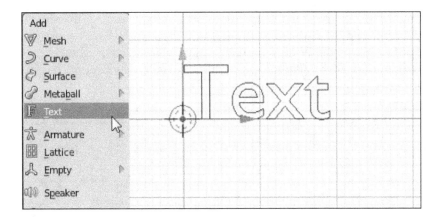

1. With a text object **Edit Mode** (*Tab*) changes to a text edit mode, similar to a word processor, instead of editing the points, edges, or faces. The tools on the left-side bar change to be text related as well.

2. In edit mode *Backspace* to clear the default **Text** and type "1 2 3 4 5 6" (The numbers 1 through 6 with a space between each). Then rename this object to Numbers. Then exit **Edit Mode** (*Tab*).

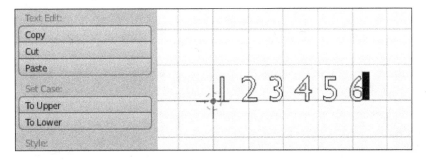

3. On the Properties panel on the right-side there is a new button option for text that has an icon like the letter **F**. This will allow the manipulation of the text object in ways suitable for 3D, all while retaining the ability to edit the text.

4. Adjust the view so that the **Spinner** and **BoxExterior** and **BoxInterior** are all visible so that the text can be modified in context. In the **Properties** panel, **Font** tab, locate the **Font** section. Change the font size to about 15. Scroll down and locate the **Paragraph** section and select the **Center** button (under **Align**) so the numbers are nicely oriented.

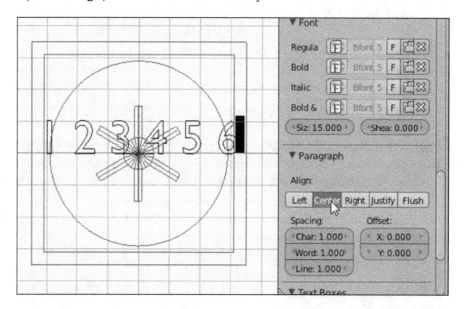

5. Higher on the **Font** tab locate the **Geometry** section. Change the **Extrude** setting to 1 to give the numbers some depth.

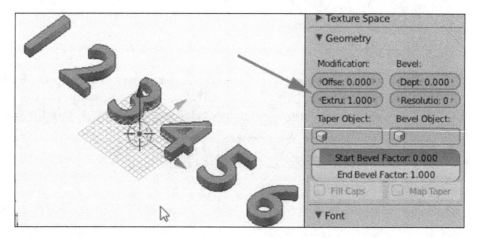

6. **Move** (*G*) the **Numbers** along the z axis (*Z*) 4 units so they sit half-way in the surface of the spinner, at the right depth to be differenced out when the time comes.

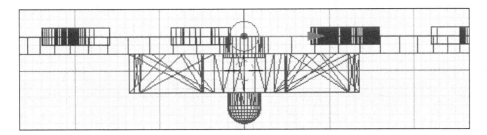

7. But that time has not yet arrived because the numbers need to be put in a circle around the wheel. First the circle needs to be added so that the numbers will be curved around. **Add** (*Shift + A*) a **Curve | Circle**.

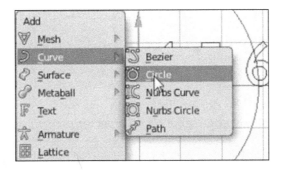

There is a significant difference between curves and meshes. With a mesh any curve perceived is the result of having a lot of straight lines in a small space, but zoom in close enough and the illusion is broken. With curves on the other hand the lines are mathematically defined, and thus remain smooth at all zoom levels. Curves can be used to do modeling if a curve surface is calculated, but curve modeling cannot take advantage of many modifiers, such as Boolean that make modeling easier without first being converted into a mesh. Additionally when exported to an STL some resolution will have to be decided on since STLs do not use curves.

8. The **BezierCircle** may not be obvious, but it is there in the middle with a radius of 1 (diameter of 2). Select (*right-click*) the numbers and in the font menu in the **Properties** panel locate **Text on Curve** (under the **Font** section). Click on the option and select **BezierCircle** from the menu.

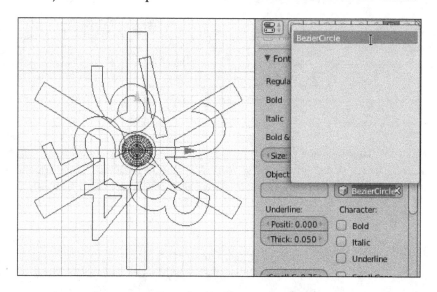

9. Select (*right-click*) the **BezierCircle** again and **Scale** (*S*) it up until the numbers are near the outside edge (about a scale factor of 17).

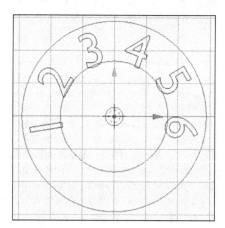

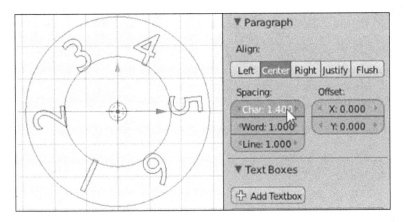

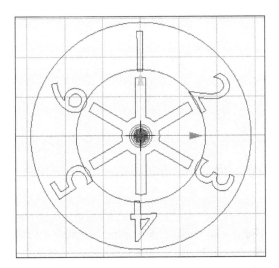

10. Select (*right-click*) the **Numbers** object locate the **Spacing** options in the **Font** menu (under **Paragraph**), and adjust the **Char** spacing until the numbers are evenly distributed around the spinner (about 1.4).

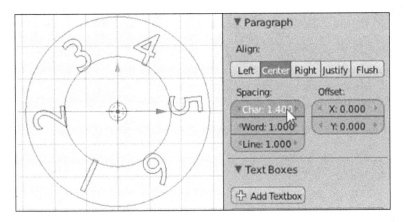

11. Finally **Rotate** (*R*) the numbers until they are lined up with the spokes of the gear.

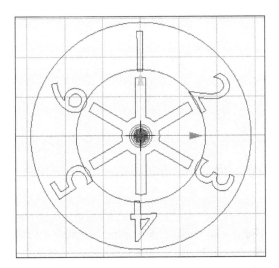

If an attempt is made at this point to **Boolean Difference** the **Numbers** from the **Spinner** it will be discovered that the **Numbers** are not available on the list of objects that can be used. This is because the numbers are defined as curves and Boolean only works with meshes. The numbers must first be converted to a mesh but in doing so they will become uneditable except as a mesh with points and lines and faces.

1. In case a future edit is necessary, select (*right-click*) the **Numbers** and **Duplicate** (*Shift + D*) them. Press *Enter* without moving the mouse to create the duplicate on top of the original. Convert the duplicate to a mesh by pressing *Alt + C* and selecting **Mesh from Curve/Meta/Surf/Text**.

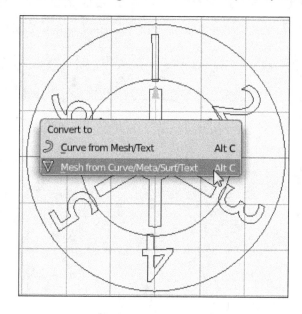

2. Then select (*right-click*) the Spinner and add a **Boolean** modifier to **Difference** the **Numbers.001** object from the spinner.

Because the numbers were converted from a curve object, they are complex, so this operation might take quite a while, and Blender may be unresponsive while calculating and from time to time after this. If it ever needs to recalculate this operation, which will be frequent. To prevent this it is best; once the operation completes the first time, and it's known that the operation will work, to turn off real-time rendering of this modifier by clicking on the eye icon in this modifier's settings. This means that while the modifier will still be on the stack it will not be active until it is turned back on.

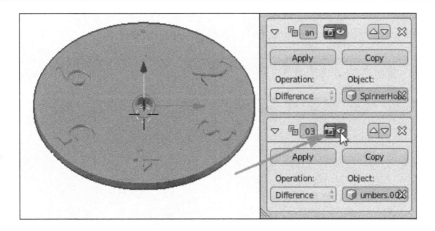

Building a rack

There needs to be something inside the spinner that will flick the gear teeth to make the spinner spin. This will be accomplished with a part that will slide past the spinner with a finger sticking out to catch the gear teeth as it goes by. The spring attaches to this part and returns it to its resting position when released to catch the gear after spinning, and hold the numbers still. This piece could extend from the back of the spinner as a simple button that is pressed to accomplish the spin, but to make things more stylish this piece will curve around the bottom of the spinner box with a finger hole.

1. Begin by revealing the **SpringExtended** and **SpringCompressed** objects created earlier and rotating (*R*) them 90 degrees around the x axis (*X*). Move (*G*) them about 23 along the x axis (*X*) and move (*G*) them again along the y axis until they are both resting against the bottom of the box. This doesn't have to be exact, it's just a guide for future building.

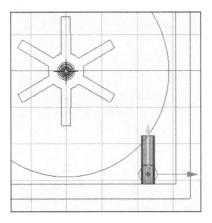

2. **Add** (*Shift + A*) a cube, and using the **Properties** sidebar (*N*) set its location and dimensions according to the following illustration:

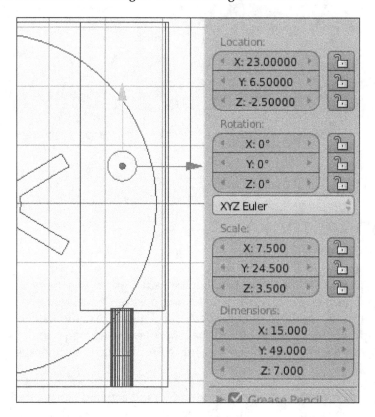

3. Name this cube Rack.

A rack is a sort of straight-line gear. This gear will only have 1 tooth in the end so calling it a rack may be a bit of a stretch, but it needs a name.

4. Enter **Edit Mode** (*Tab*). Loop cut (*Ctrl + R*) the rack and place the new points at about one-third and two-thirds of the way between the gear teeth. **Border Select** (*B*) the points on the left hand side and **Extrude** (*E*) about 5 units out to make the tooth that will spin the gears.

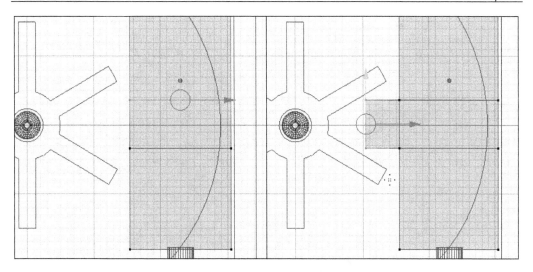

5. Exit **Edit Mode** (*Tab*) and move (**G**) the **Rack** along the y axis (**Y**) until it rests at the point where the compressed spring ends to simulate depressing the button to spin the **Spinner**. The **Spinner** won't actually react but this will provide some important information.

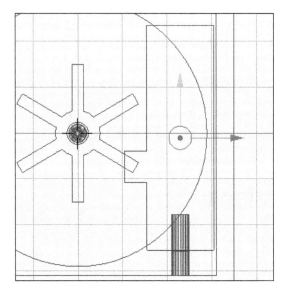

6. It is clear that with the rack depressed the spinner cannot spin freely, the tooth gets in the way. Re-enter **Edit Mode** (*Tab*) and modify the tooth so it is out of the way while the spinner spins.

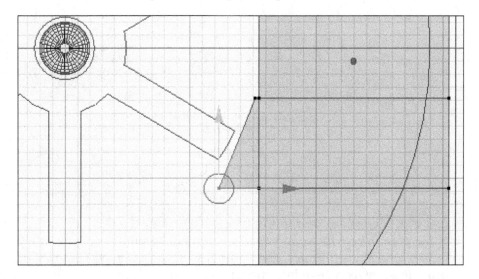

7. Out of **Edit Mode** (*Tab*) the modifications can be tested by **Rotating** (*R*) the **Spinner** being sure to **Undo** (*Ctrl + Z*) the rotation when testing is complete. The **Spinner's** gear teeth should be able to rotate freely without touching any part of the **Rack**.

8. Then move (*G*) the rack back up along the y axis (*Y*) until it rests where the extended spring sits, and test to be sure that the gears are intercepted when the rack is in this position. There will, naturally, be some wiggle but for the most part it should hold one number still for long enough for the number to be viewable.

Adding the trigger

1. To add the switch so a finger can spin the spinner in **Edit Mode** (*Tab*) select the top vertices, move (*G*) them along the y axis (*Y*) 6 units then **Extrude** (*E*) them 6 units.

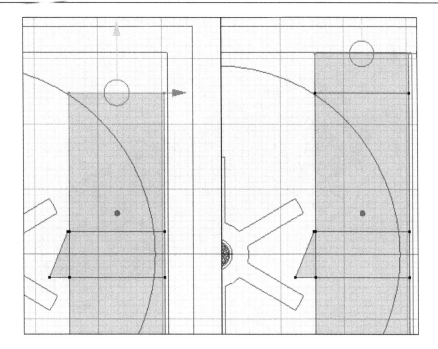

 When possible this method of moving then extruding can be a more accurate way to do geometrically similar operations to a loop cut.

2. Loop cut the body of the rack in half, vertically. Then from a view looking up from below, select the face in the corner of the bottom. (This can be accomplished simply by selecting the 4 vertices that comprise the desired face.)

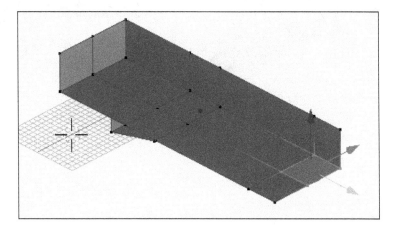

3. **Extrude** (*E*) this face 3 units, then **Extrude** (*E*) again 4 units.

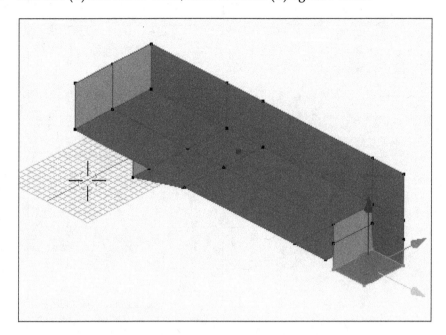

4. Deselect (*A*) the current face, then select the face on the front of the newly extruded arm, around the corner from the last selected face. **Extrude** (*E*) this face 24 units.

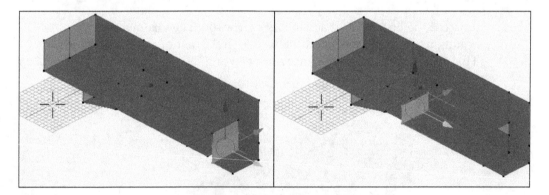

5. Deselect (*A*) the current face then select the face on the bottom of the newly extruded arm, around the corner from the last selected face again. **Extrude** (*E*) this face 20 units.

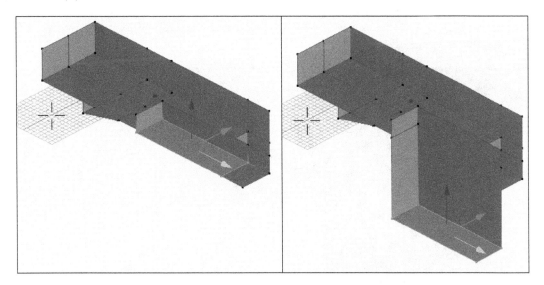

A significant amount of planning goes into making anything for 3D printing, particularly when printing without supports. It may not be obvious, but the spinner was designed to be printed on its face and the trigger can be turned 90 degrees and printed on its side with the tooth sticking up.

6. With the basic shape of the rack in place a hole for the finger needs to be made. Exit **Edit Mode** (*Tab*) and **Add** (*Shift + A*) a cylinder. Name it RackFingerHole.

7. In the **Properties** (*N*) sidebar set the dimensions to **X:** 20, **Y:** 20, **Z:** 15, set the location to **X:** 27, **Y:** 13, **Z:** -21, and set the rotation to **X:** 0, **Y:** 90, **Z:** 0. (Some adjustment in the y axis may be necessary after placement based on where the rack may be sitting after the simulated presses. Try to make the `RackFingerHole` so it is approximately 2mm from the edge of the rack's finger space all around.)

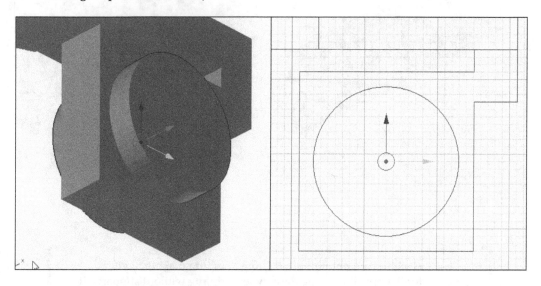

8. Add a **Boolean** modifier to the rack to **Difference** the `RackFingerHole` from it.

Docking the spring

The rack is almost complete with the exception of something to hold the spring in place. For this a small protrusion small enough for the spring to go around should do. Refer back to the measurements of the spring taken at the beginning. The inside of the spring was approximately 3mm in diameter.

1. This connection should be snug so **Add** (*Shift* + *A*) a cylinder with a radius of 1.5 and a depth of 3. Name it `RackPeg`. **Move** (*G*) it 23 along the x axis (*X*) and move (*G*) it along the y axis (*Y*) until it sticks just below the bottom of the rack, or about -18 if the rack is exactly where it started.

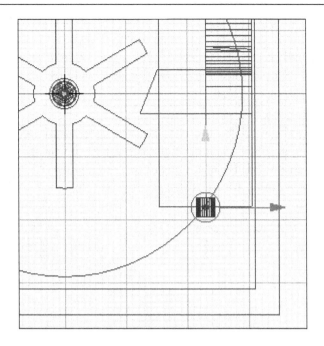

2. Switch to **Front** view (*Numpad 1*) and move (*G*) the **RackPeg** along the z axis (*Z*) 2.5.

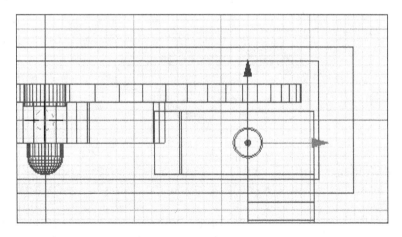

3. Select the **Rack** and **Boolean** Union the **RackPeg** into it.

Doesn't the peg violate the 45 degree rule? Unless this object is printed with the peg sticking up, the 45 degree rule is going to be violated, and if it's printed up then it will be violated elsewhere on this object. It seems that there is no escaping that this peg violated the rule set out in *Chapter 1, Design Tools and Basics*.

It's true that this peg does, strictly speaking, violate that rule however it is a very minor violation of it. Also, it will allow any sufficiently well calibrated 3D printer to mess up, very slightly, when printing it, build on that mess and correct itself before too many layers pass. If when the final print is done, it's found that one side of this peg is too messy than one corner it can be cut off, cleaned up, and there will still be enough of the peg remaining for the spring to hold on to.

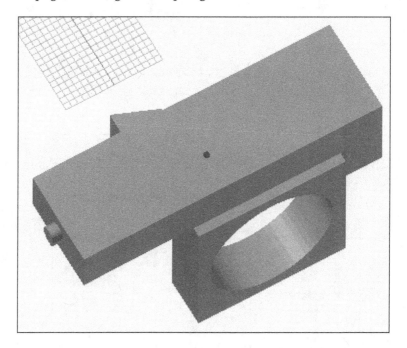

Make sure the other objects are visible, and take a moment to check in wireframe mode (Z) to be sure that the rack is sitting below the main disk of the **Spinner** and not intersecting with it. Also, the main body of the rack should be entirely inside the **BoxInterior**. If the view is adjusted the finger hole should extend outside the body but only at one small part and the majority of the finger hole should be outside the **BoxExterior**.

Does the location in the virtual space really matter? When designing an object that will be separated, reoriented, printed in multiple parts its location in relation to other objects doesn't really matter. As long as the dimensions are correct according to a well laid out plan the result will be the same. However, keeping the objects in relation to each other in virtual space does have its advantages. No plan is perfect and troubleshooting potential problems before printing is easier if all the objects are aligned. Fixing these problems is also easier. Does this catch and fix all problems and ensure prints will be perfect? No, but it does eliminate one chance for error.

Modeling the case – lid

The case will be split into two parts. The lid will contain a window for viewing the number and a peg to hold the spinner in place. The bottom will have a place for the spinner's peg to rest, something to hold the rack in place securely, and a slot for the rack to slide.

First a box will be defined that will show the space that the lid will occupy, which can be used for both the top and bottom.

1. **Add** (*Shift* + *A*) a cube. Name it LidSpace and change its **Location** and **Dimensions** in the **Properties** (*N*) sidebar according to the following illustration:

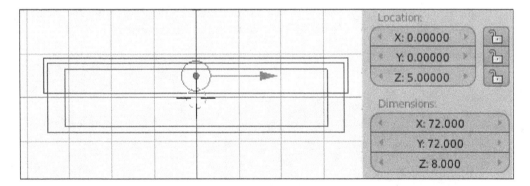

2. **Add** (*Shift + A*) another cube. Name it LidLip and change its **Location** and **Dimensions** accordingly:

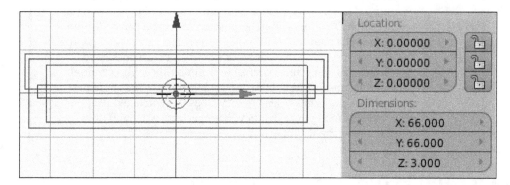

3. Select (*right-click*) the **LidSpace** and **Boolean** union the **LidLip** to it.

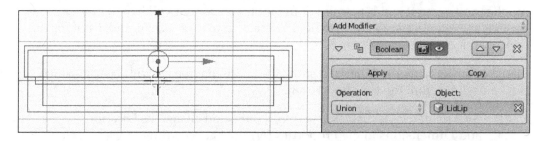

4. Select (*right-click*) the **BoxExterior**, duplicate (*Ctrl + D*) it and name the duplicate BoxLid. On the **BoxLid** add modifiers to **Boolean Intersect** with the **LidSpace** and **Boolean Difference** with the **BoxInterior**.

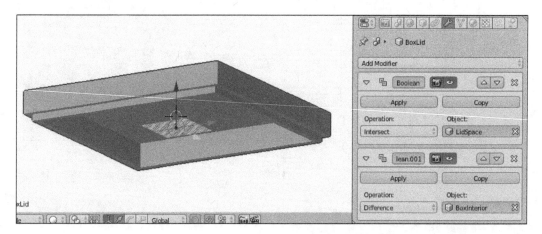

5. To make a window for viewing the numbers, **Add** (*Shift* + *A*) a **Cylinder**. In the **Properties** (*N*) sidebar change its **Dimensions** to **X:** 54, **Y:** 54, **Z:** 5. Move (*G*) it up 7 along the z axis (*Z*) so it intersects the **BoxLid**. Name this object ViewWindow.

6. **Add** (*Shift* + *A*) another **Cylinder** and change its **Dimensions** to **X:** 30, **Y:** 30, **Z:** 6. Move (*G*) it up 7 along the z axis (*Z*) too so it intersects the **ViewingWindow** and **BoxLid**. Name this object ViewHole.

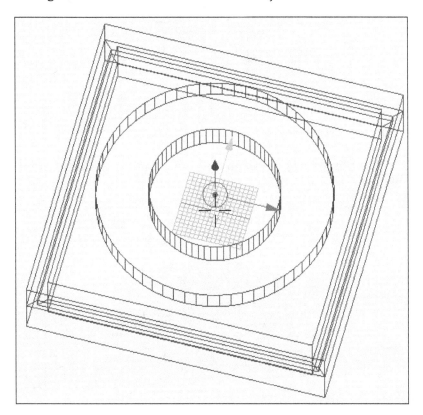

7. Select (*right-click*) the **ViewWindow** and **Boolean Difference** the **ViewHole** from it.

8. **Add** (*Shift + A*) a cube. Rename the cube Viewwedge. Move (*G*) it along the z axis (*Z*) by 7 and along the y-axis (*Y*) by 20. **Scale** (*S*) along the z axis (*Z*) by 3. **Scale** (*S*) it in the x and y axis (*Shift + Z*) by 8.

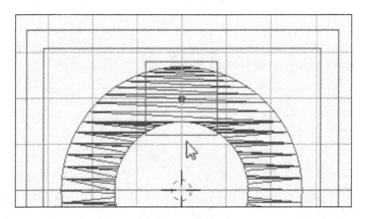

9. Enter **Edit Mode** (*Tab*). Select all the points on the right and **Rotate** (*R*) them 30 degrees.

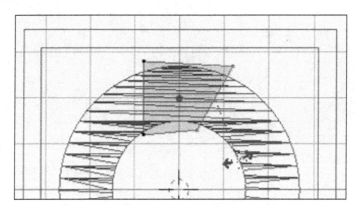

10. Select all the points on the left side and **Rotate** (*R*) by -30 degrees.

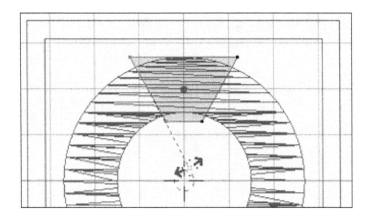

11. Select (*right-click*) the **ViewWindow** again and **Boolean Intersect** it with the **ViewWedge**.

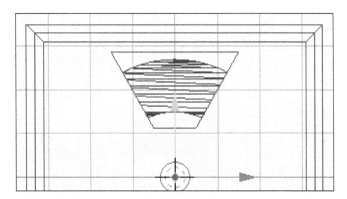

12. The select the **BoxLid** and **Boolean Difference** the **ViewWindow** from it.

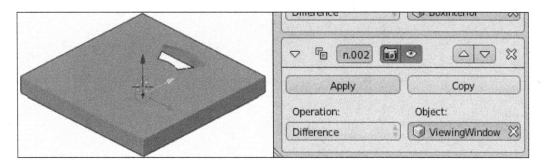

13. It is probably helpful at this point to be sure that the **BoxExterior, BoxInterior, BoxLid, Spinner** and **SpinnerPeg** are all visible. Select (*right-click*) the **SpinnerPeg**. **Duplicate** (*Shift + D*) it and move the duplicate along the z axis (*Z*) 8 units. Name the duplicate Lidpeg.

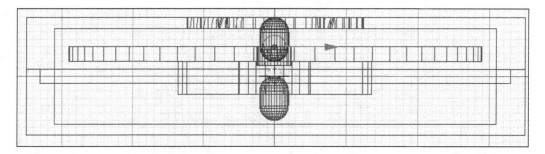

14. The top of the **LidPeg** doesn't quite fit inside the wall of the **BoxLid** entirely so in **Edit Mode** (*Tab*) select all the points on the top and scale them by 0 along the z axis (*Z*) to flatten it. All the points should be inside the top wall of the **BoxLid**.

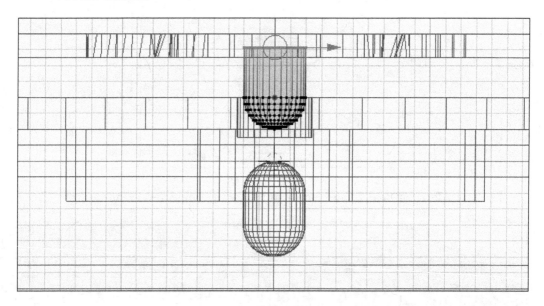

15. Select the **BoxLid** and **Boolean Union** the **LidPeg** to it.

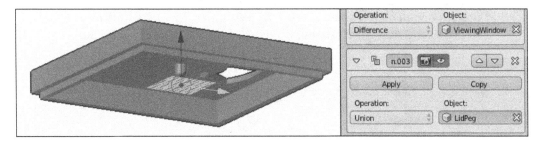

Modeling the case – bottom

1. Select (*right-click*) the **BoxExterior**, **Duplicate** (*Shift + D*) it and rename the duplicate BoxBottom. **Boolean Difference** the **BoxInterior** and **Difference** the **LidSpace** from the **BoxBottom**.

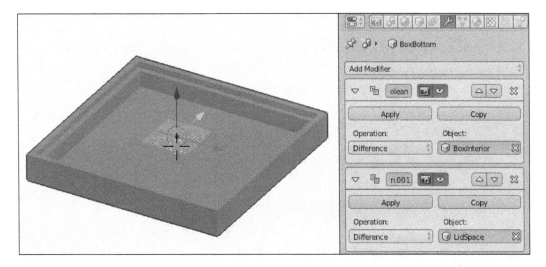

2. Like before, having a few other objects visible for reference will help with this step. Be sure that the **BoxExterior, BoxInterior, BoxBottom** and **Spinner** at least are all visible. **Add** (*Shift + A*) a cylinder. Make its **Radius** 2.4, its depth 5, and move it along the z axis -6. Name it BoxBottomPegHole.

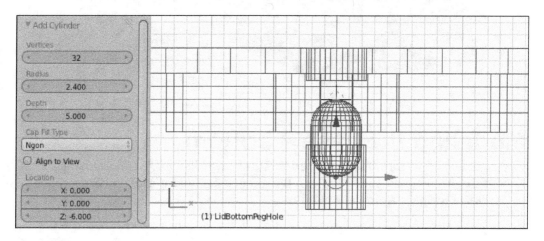

3. **Add** (*Shift + A*) another cylinder with **Radius** 6, **Depth** 3, and move it along the z axis -6. Name it BoxBottomPegHolder.

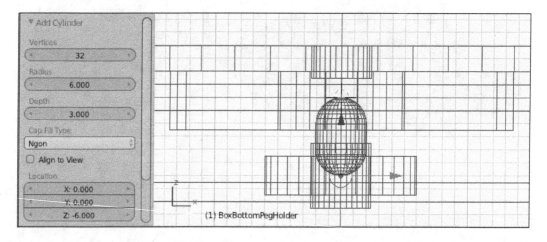

4. **Boolean Difference** the **BoxPegHole** from the **BoxBottomPegHolder**. Then select (*right-click*) the **BoxBottom** and **Boolean Union** the **BoxBottomPegHolder** to it. Be careful to do these steps correctly or there may end up a big hole through the bottom of the **BoxBottom**.

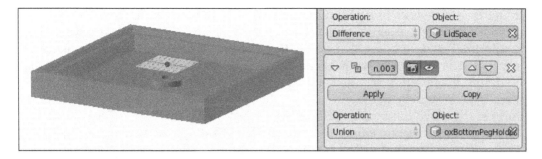

Make the **Rack** visible. As the rack slides up and down it is possible that it might wiggle left and right, getting in the way of the gear and stopping the **Spinner**. It is therefore necessary to add something to keep the rack in place.

5. **Add** (*Shift + A*) a cube, name it **BoxBottomRackGuide** and change its location and dimension accordingly:

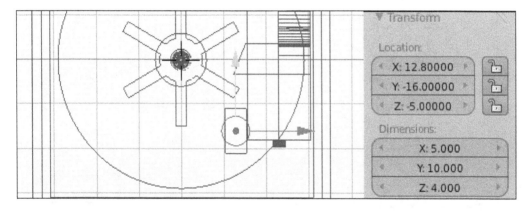

6. Select (*right-click*) the **BoxBottom** and **Boolean Union** it with the **BoxBottomRackGuide**.

Many of the illustrations in this book employ tricks to simplify what's seen on the screen. This is done for clarity, but it has masked the fact that when using the Boolean modifier, the screen can be filled up with extra geometry that can clutter the view. For the most part this extra geometry can be ignored however, sometimes simplifying this geometry is a desirable goal.

In this case there is an opportunity to simplify the geometry of the **BoxBottom** using a very simple trick. With the order of the Boolean operations as it is when the **BoxBottomPegHolder** is added extra edges must be created from the inside corner to the various points where the cylinder intersects the inside bottom. Then the **BoxBottomRackGuide** is added and it cuts through most of those extra edges, but there is no attempt by Blender to remove the superfluous geometry. However, if the order of these two Boolean operations is switched, so the **BoxBottomRackGuide** is done first, then the extra edges aren't there when it's added, and the extra edges for the **BoxBottomPegHolder** starts at the corner of the **BoxBottomRackGuide**.

This has absolutely no impact on the object's printability. Occasionally, complex geometry can cause Boolean operations to fail so a simpler geometry might be necessary in those cases. But for now the only difference is that it looks better when editing. Do not feel any obligation to worry about this step.

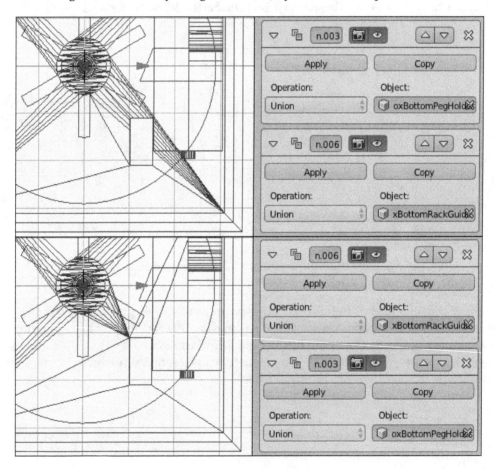

1. Make visible the **RackPeg**, select (*right-click*) it, and **Duplicate** (*Shift + D*) it. Move the duplicate along the y axis (*Y*) until it is sticking out the inner wall of the **BoxBottom**, about `-13` (depending on where the rack was). Name the duplicate `BoxBottomSpringPeg`.

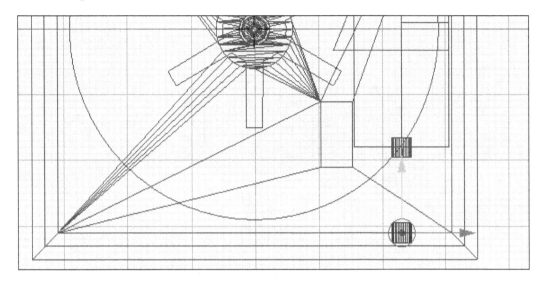

2. Add another **Boolean** modifier to the **BoxBottom** and **Union** the **BoxBottomSpringPeg** to it.

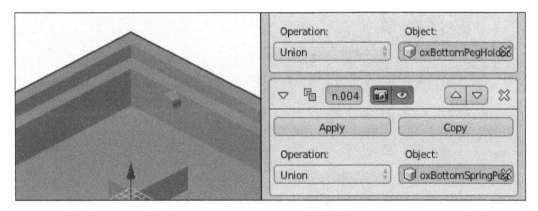

3. Lastly, a slot needs to be made in the **BoxBottom** for the **Rack** to slide. The slot needs to be long enough to accommodate the extension of the spring.

4. **Add** (*Shift + A*) a cube. In the **Properties** (*N*) modify its **Location** and **Dimensions** as in the illustration and name it `BoxBottomSlot`.

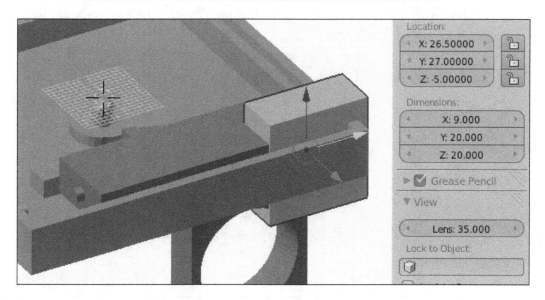

5. Select (*right-click*) the **BoxBottom** and **Boolean Difference** the **BoxBottomSlot** from it.

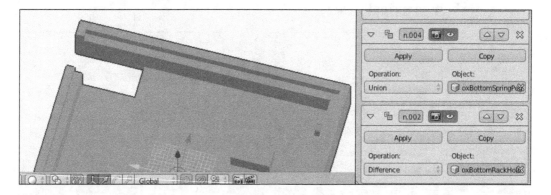

Preparing for print

Finally all the parts are ready, and all that's left is to orient them for print, similar to the way it was accomplished for the modular toy robot.

1. Select (*right-click*) the **BoxBottom**, **Duplicate** (*Shift + D*) it, name the duplicate `PrintBoxBottom`, and view in **Local View** (*Numpad /*) apply all modifiers in order. This part is ready to print as oriented so **File | Export | STL**, and name the exported file `SpinnerBottom.Stl`. Then **Hide** (*H*) **PrintBoxBottom** and exit **Local View** (*Numpad /*).

2. Select (*right-click*) **BoxLid**, (make it visible if it is not). **Duplicate** (*Shift + D*) it, name the duplicate `PrintBoxLid`, and view in **Local View** (*Numpad /*). Apply all modifiers in order. **Rotate** (*R*) `180` around the y axis (*Y*). **File | Export | STL** and name the exported file `SpinnerLid.Stl`. Then **Hide** (*H*) **PrintBoxLid** and exit **Local View** (*Numpad /*).

3. Select (*right-click*) the rack, (make it visible if it is not). **Duplicate** (*Shift + D*) it, name the duplicate `PrintRack`, and in **Local View** (*Numpad /*) apply all modifiers in order. **Rotate** (*R*) `90` around the y axis (*Y*). **File | Export | STL** and name the exported file `Spinnerrack.Stl`. Then Hide (*H*) **PrintRack** and exit **Local View** (*Numpad /*).

4. Finally select (*right-click*) spinner, (make it visible if it is not). **Duplicate** (*Shift + D*) it, name the duplicate `PrintSpinner` and view in **Local View** (*Numpad /*) apply all modifiers in order. Remember with this one the last modifier for the letters may take a while. Be patient because once it's applied it won't cause any more pauses. **Rotate** (*R*) `180` around the y axis (*Y*). **File | Export | STL** and name the exported file `Spinner.Stl`. Then **Hide** (*H*) **PrintSpinner** and exit **Local View** (*Numpad /*).

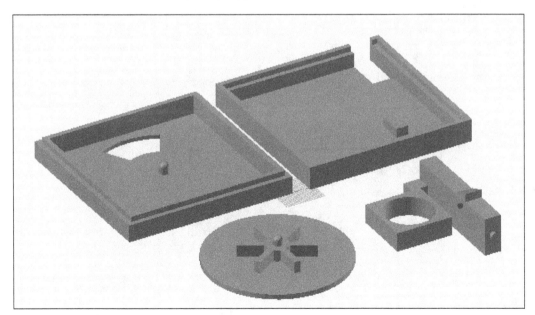

Printing and assembly

1. Print `Spinner.stl`, `SpinnerLid.stl`, `SpinnerBottom.stl` and `SpinnerRack.stl`. Also have the spring ready.

2. Take the rack and carefully wedge it around the bottom at its slot. There may be some flexing necessary to get it to work but be gentle.

3. Place the spring around the peg on the base and as the rack gets pushed in around the peg at the base of the rack. Test the connection a few times by compressing and releasing the rack.

4. Place the spinner on top of the rack mating the spinner's peg into the holder for it in the bottom.

5. Firmly attach the lid, being sure that the lip goes in to the bottom all the way around. The fit will be tight which should prevent the need to use any glue to hold the parts together.

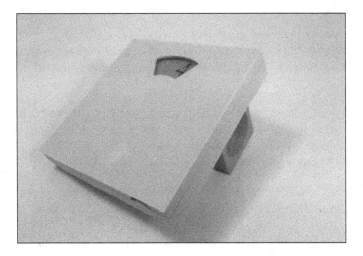

At last the spinner is ready to use to generate random numbers.

Extra credit

Is having the numbers in sequence the best idea or should they be scrambled on the wheel? When making dice, the pips are placed in such a way that they balance themselves. Balance isn't a problem with this spinner, but perhaps it would "feel" better, if they were in a more random order. And why do they have to be numbers? Pips or even pictograms could be used instead.

Could the spring be replaced with something that is printed in 3D? Plastic is flexible and printed springs have already been proven effective. Perhaps one of them could be incorporated into this project?

This project could be extended in other ways. A D6 roller is fine, but how would a different number of options work out? The gear is designed to stop the numbers in specific places. Eight numbers and hence eight teeth on the gears would probably work fine, but how could a D20 be implemented?

What about a 2D6, spinning two disks simultaneously and outputting two different numbers. Perhaps the finger switch underneath would have to be abandoned, but more than that how could both spinners be spun at slightly different rates so they're not just spinning in tandem, always returning the same pairs, or something close to them? The answer to this might be to make gears slightly different sizes, and have them interact with the teeth at different radial depths. How would this be implemented? Is there a different way to spin both spinners at different rates?

Summary

In this chapter, we came across gears from an add-on in Blender, parts designed to move freely within constraints, and a print inserted to create motion. This has been an interesting project. Also, hopefully its application to useful devices should be obvious.

While Blender's gear add-on may not be the best, the gears it produces can be perfectly serviceable. There are other gears that can be downloaded from Thingiverse, if something of a more technical sort is desired, which can then be imported into Blender, and incorporated into any project.

The creation of mechanical interactions is never easy, but it can be one of the best uses for 3D printers. Blender has proven to be adequate at this task.

In the next chapter, new modeling tools will be introduced and explored that will open up new modeling possibilities. Skinning a mesh and sculpting tools will be combined to make a detailed, organic model.

7
Teddy Bear Figurine

Basic shapes, vector manipulation, and Boolean modifiers aren't the only way to model in Blender. One of Blender's strengths is the eclectic blend of tools that it offers. In this project a simple organic shape will be modeled and posed for printing. Starting with a basic stick figure, then fleshing the stick figure out and giving it proper form and shape. And that shape will be of a Teddy bear!

The rules of 3D modeling like overhang and bridges are still important, even if the tools that were used have changed. But for the most part, those rules can wait until the end where they will be applied, with a little finesse.

Making a stick figure

Begin this project like all the rest. Clear the scene and save it in a new directory under **MakerbotBlueprints** called Ch 7 Teddy Bear and name the file Teddybear.Blend.

In the beginning all that is needed is a single point. Unfortunately Blender doesn't have an object that is just a single vertex. Fortunately it does have many basic shapes from where a single point can be taken.

1. Begin by adding (*Shift* + *A*) a **Plane**.

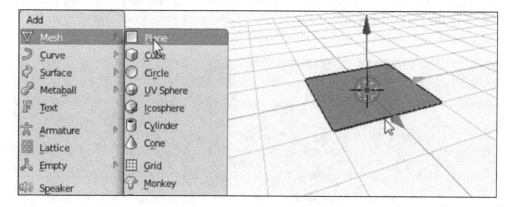

2. Rename this plane Bearskin.
3. Enter **Edit Mode** (*Tab*).
4. Select all but one vertex and **Delete** (*X*) them so there is only one left.
5. Select and **Move** (*G*) the one remaining vertex to the origin.

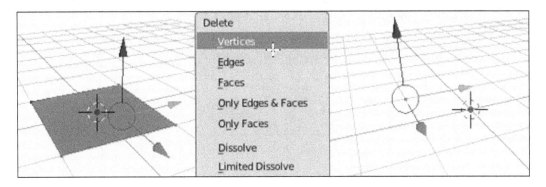

 It is important that this point ends up exactly at the origin for the next step to work. With a plane whatever point is left is exactly 1 unit in two axes away from the origin, so the best way to ensure it is at the exact origin; this is where we should type the commands and not use the mouse. For the earlier illustration, typing the following keys will center the point perfectly: *G, X, 1, G, Y, 1.*

6. Jump to front **Orthographic** view (*Numpad 1, Numpad 5*).

7. **Extrude** (*E*) this point along the z axis (*Z*) approximately 10 units. The exact distance doesn't matter but locking this point to the z axis is important. This line will form the body of the stick figure.

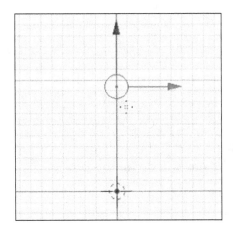

8. **Extrude** (*E*) again along the z axis (*Z*) about 6 units to form the neck.

9. Then select (*right-click*) the point at the base of the neck and **Extrude** (*E*) again, this time moving freely to position a shoulder.

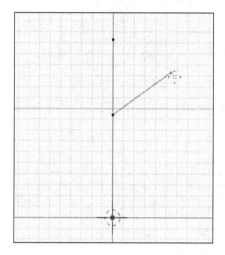

10. From the shoulder **Extrude** (*E*) to an elbow and again to a wrist.

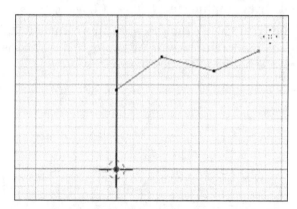

11. From the wrist **Extrude** (*E*) a thumb and palm. Don't bother with all the fingers, just one big one will do for this project.

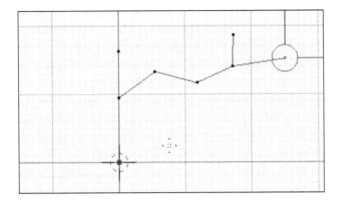

12. Select (*right-click*) the point at the origin. This will be the tummy point.

13. **Extrude** (*E*) a hip, thigh, calf, and a little extrusion for the foot. Don't worry too much about placement at this point, since it will be adjusted later when there's skin on the bones.

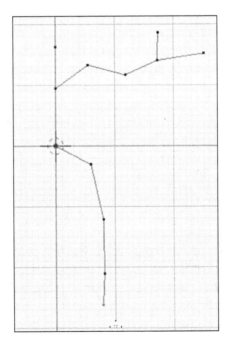

Finally form the head:

1. Select (*right-click*) the point at the top where the neck is present.
2. **Extrude** (*E*) a long stick along the z axis (*Z*) for the head about 10 or 15 units.

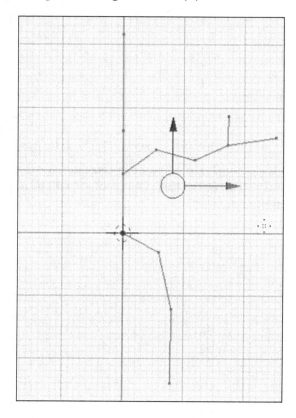

Instead of repeating the process for the other side of the body let the **Mirror** modifier do the hard work.

1. In the modifiers tab add a **Mirror** modifier.
2. In the modifier panel click on the button that looks like a triangle with the vertices highlighted.

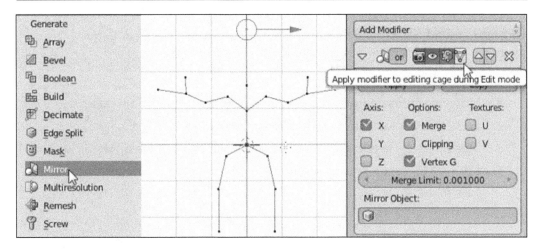

That button will show the effects of the modifier even in the edit mode. This is sometimes undesirable for modifiers that will slow down the computer, but in this case it is a very good idea to turn that option on.

Putting the skin on the bones

This stick figure will be the bones of the figure. Blender has a very cool modifier that will thicken any stick mesh into a full bodied mesh. It is called, appropriately enough, **Skin**.

- Add the **Skin** modifier to the mesh

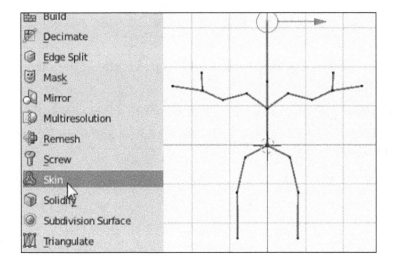

It might not be immediately apparent, but now every stick has a 3-dimensional shape wrapped around it. To make the effect more apparent, use the **Skin Resize** (*Ctrl + A*) operator. This will be a temporary step to see the effect first:

1. Select all points (*A*).
2. Move the mouse pointer to a point closer to the center of the skeleton.
3. Press *Ctrl + A* to begin the **Skin Resize** operation.
4. Move the mouse pointer away from the skeleton until the mesh thickens up.
5. *Left-click* to complete the operation.

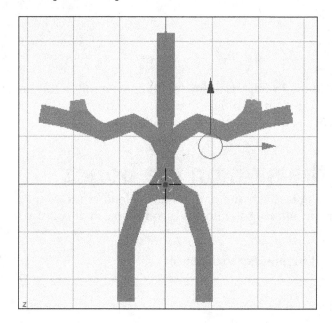

The skin resize operator can alter each point independently so the shape does not have to be uniform.

1. **Undo** (*Ctrl + Z*) the previous operation back to a thin sticks all around.
2. Select (*right-click*) the vertex where the legs meet and **Skin Resize** (*Ctrl + A*) to make a nice big belly.

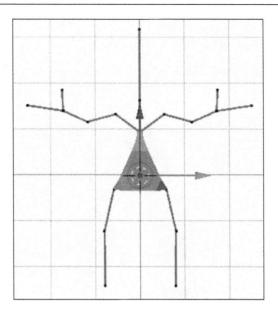

3. Select (*right-click*) the hip joint.

4. **Skin Resize** (*Ctrl + A*) it to add some volume to that joint.

Since the mirror modifier is still active changing one side will change the other automatically.

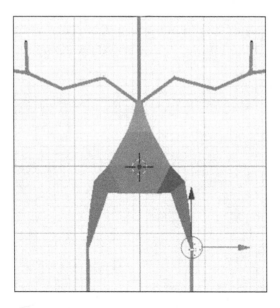

Even though the points are hidden behind the skin, they are still selectable. **Wireframe** view (Z) will make them visible if desired.

- Select (*right-click*) and **Skin Resize** (*Ctrl + A*) the knees and feet giving the teddy bear big broad feet to stand on

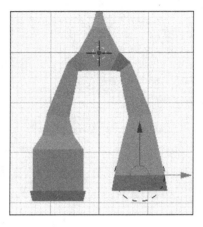

The skin plug-in has built into it a concept of a root vertex, or a vertex from which the object flows out. If this root is not set properly, it can cause problems, particularly with symmetrical objects, such as the mirrored skeleton. This is why the feet are not uniform. The root vertex is the one with a circle around it, in this case the foot. As a general rule it is a good idea to be aware of the root and choose a central vertex for it.

1. Select (*right-click*) the vertex at the origin, the belly vertex.
2. In the skin modifier setting click on the **Mark Root** button.

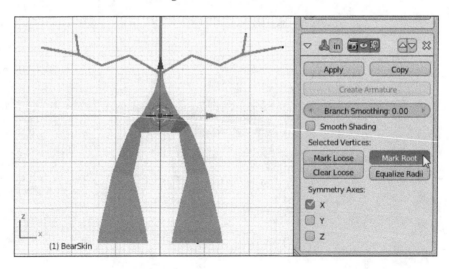

Now the skin is uniform on both sides. Continue fleshing out the skin.

1. Select (*right-click*) and **Skin Resize** (*Ctrl + A*) the chest vertex, shoulder vertex and elbow vertex.

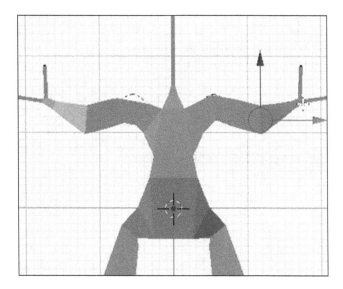

2. Select (*right-click*) and **Skin Resize** (*Ctrl + A*) the wrist.

    It is possible when skin resizing on the wrist to have the wrist entirely envelope the thumb. If this is the case simply move (*G*) the thumb vertex so it continues to stick out from the palm.

3. Select (*right-click*) and **Skin Resize** (*Ctrl + A*) the finger tips to finish the hand shape.

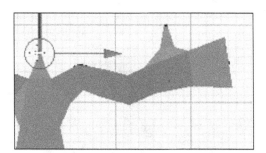

At this point it becomes apparent that the joint of the shoulders would be better, if it weren't included in the skin calculations, so the skin went from shoulder to shoulder. The skin plugin has a function for this.

1. Select (*right-click*) the vertex where the shoulders join to the body.

2. Click on the **Mark Loose** button in the skin modifier properties, and the skin will go from shoulder-to-shoulder without trying to fit around that point making a more natural body shape.

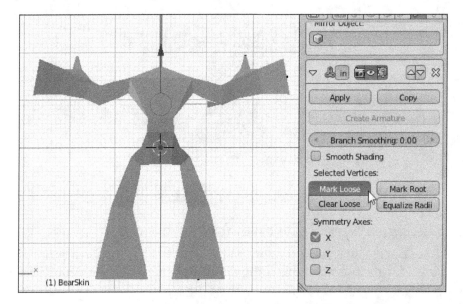

3. Select (*right-click*) and **Skin Resize** (*Ctrl + A*) the neck to give the head the suggestion of a bulbous shape.

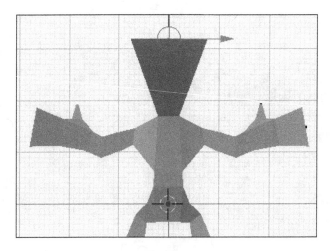

Smoothing the skin

At this point things are a little too blocky. It's time to smooth these rough edges. In past projects the **Multiresolution** modifier was applied to smooth things out. In this project that is not an option, because the Multiresolution modifier cannot be applied after the skin and mirror modifier on the modifier stack. Instead use the Subdivision Surface modifier, a similar modifier to the Multiresolution modifier without some of the restrictions and functionality.

- Add a **Subdivison Surface** modifier
- Change the settings in the modifier's properties to three subdivisions for view and render

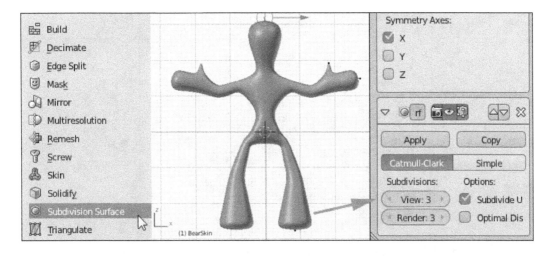

View and render in subdivision and Multiresolution refer to what is seen in the edit panel (view) and what is used when making a 3D render of an object. The idea is that while working the object can be low resolution so the computer isn't over taxed, and the resolution can be turned up when looks matter. But when exporting the STL it is the view setting that gets applied to the mesh.

If Blender is only being used for modeling for 3D printing does the render setting even matter? Not really. However it is good practice not to ignore a setting when it is that close to the important one, just in case.

What comes next is officially known as "fiddling". To this point the model was intentionally made different from the final shape it will take. This is how 3D modeling is often done; less science, more art.

Take some time and move (*G*) vertices around, re-skin resize (*Ctrl + A*) them, adjust, push, prod, and generally mess with it until it looks right. Make it personal. There is no right or wrong in art.

One thing to be careful of in this process is the bad fold. Sometimes a vertex's area of influence overlaps one next to it and the program can't tell where the points should go so it makes a bad judgment and the result is a bad fold:

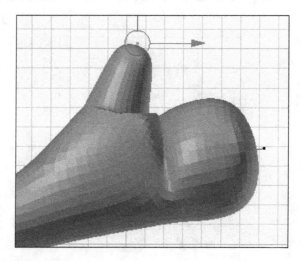

Try to avoid bad folds. There will be plenty of bad geometry generated later when the model is posed, it is best to avoid it here. Make sure at this point that the mesh is nice and smooth all around. The way to fix it is simply to fiddle until it goes away. Pull vertices apart or decrease their skin resize area of influence.

Give this process some time and feel free to experiment. At the end of the fiddling process the final model can be very different from where it started.

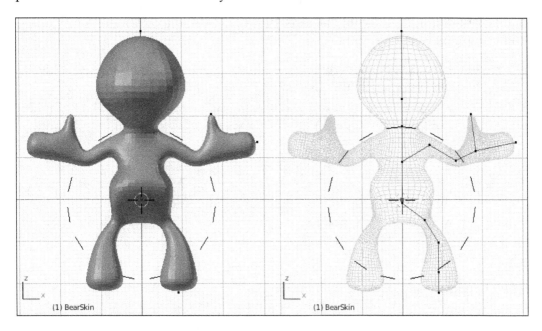

Extruding (*E*) additional vertices at the extremities allows for finer control over the contours shape, but don't deviate too much from the established skeleton. Finer details can be added later; at this point it is somewhat important that the skeleton "make sense" for posing purposes and that won't happen if the arm has three elbows.

Adjusting for the third dimension

As it is so far this is a fine gingerbread man with its stance so flat. To add some life to this model, simply adjust it in all three dimensions. And to aid in this a new view will be introduced; **Quad View**.

- In the menu at the bottom of the **3D View** panel select **View | Toggle Quad View** or press *Ctrl + Alt + Q*.

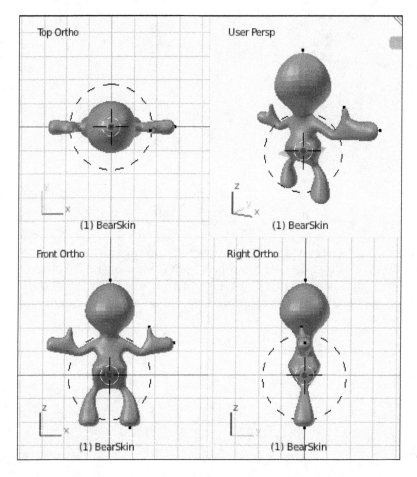

The utility of this view is to simultaneously view the model from many different angles. The disadvantage is that each of the views is a quarter the screen size making details hard to see.

In quad view the location of the pointer when using hot keys is more important than ever:

1. As long as the mouse pointer is over any of the 3D views the number pad keys will only adjust the upper-right **User Persp** view.

2. Using the *middle-click* will only free rotate the view point if the mouse pointer starts over the upper-right view and will do nothing in the other views.

3. Zooming with the mouse wheel or the *Numpad +* and *Numpad -* keys only zooms the view that the mouse pointer is over.

4. Panning the view (*Shift + middle-click*) only affects the view where the mouse pointer was when it started.

Operators like **Grab** (*G*) and **Rotate** (*R*) operate according to the view that the mouse pointer is over when the operation started. In other words if the grab operation is begun while the mouse pointer is over the **Right Ortho** view then it will be locked to the **YZ** plane. If the mouse pointer is over the **Top Ortho** view then it will be locked to the **XY** plane.

To add some dimension to the model:

1. Select (*right-click*) the shoulders and **Move** (*G*) them back slightly along the y axis.

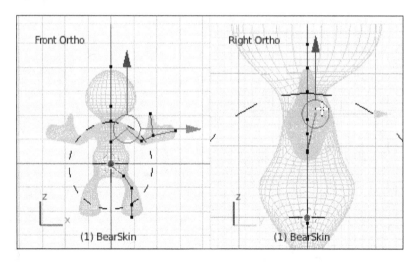

2. Select (*right-click*) the elbow and **Move** (G) it back slightly further than the shoulder moved.

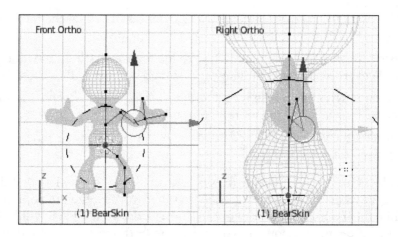

3. Select (*right-click*) the fingers and thumb together and **Move** (G) them forward.

4. The wrist can remain where it is.

    Now the arm has a more natural look to it.

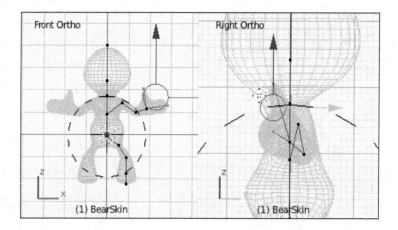

5. Select (*right-click*) and **Move** (G) the knees a little forward.

6. Select (*right-click*) and **Move** (G) and the hips a little backwards.

Notice how this affects the curve of the teddy bear's belly because the angle of the points is different.

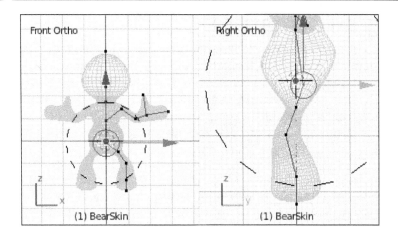

Adjusting the bear's dimensions in this way may seem like a little thing, but it goes a long way to making a more organic looking and three dimensional model.

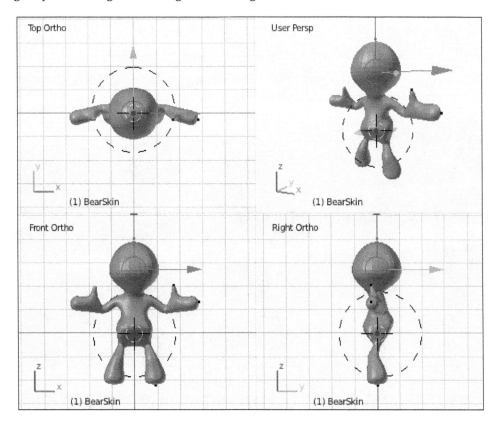

- When complete press *Ctrl + Alt + Q* to switch out of **Quad View**

Making an armature

An armature is used generally in 3D animation to move the model in natural ways. The idea and name comes from Claymation where a metal frame, sometimes with hinges sometimes just made from bendable material, would be built into the clay model like a skeleton to hold it upright and maintain poses, while the animation is accomplished frame-by-frame. In computer modeling the armature is used in a similar way to define the pose of a complex model by manipulating just a few skeleton lines.

Normally when modeling characters the process of building an armature is a complicated one involving many steps that would be outside the scope of this work. However, by using the skin modifier the process is simplified significantly.

This marks the beginning of another major section so this is a reminder about saving (*Ctrl + S*) and incremental saves.

1. Exit **Edit Mode** (*Tab*).

2. Open the modifiers panel and **Apply** the **Mirror** modifier.

3. On the **Skin** modifier click on the **Create Armature** button.

Now the original line mesh that the skin modifier was based on will be converted to a new armature object and added to the scene.

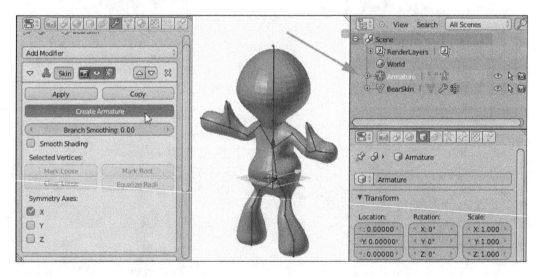

Then select the original **BearSkin** and **Apply** the **Skin** and **Subsurface** modifiers in that order, but leave the new armature modifier.

An armature is similar to a mesh in that it is a series of edges joined at points. But unlike a mesh an armature can be edited or posed. In fact armatures can have many poses saved and switch to them at will, but exploring that functionality is outside the scope of this work. Also, note that an armature, as mentioned earlier affects many points of the mesh to which it is attached.

1. Select (*right-click*) the armature.

2. Click on the **Object Data** tab on the **Properties** panel on the right. The object data for the armature is the one that looks like a little human figure.

3. This menu has the options related to the armature.

4. At the top of the first menu section, **Skeleton**, there are two buttons, **Pose Position** and **Rest Position**.

5. Be sure the **Pose Position** button is active or the skeleton won't be moveable.

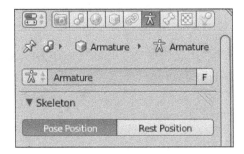

Following that are display options for the armature. The default option is **Stick** and makes the bones basically lines with dots at the joints. **Octahed** changes the bones to look like large shapes, and is preferred by some 3D animators because they're easier to see on screen. The rest of the options are basically the same for his project. If octahed is preferred use it, but the rest of this chapter will use stick in the illustrations because it shows up well in print.

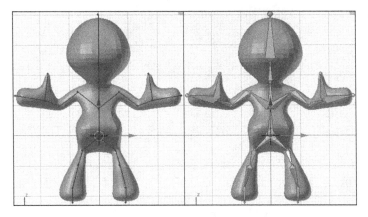

An exploration of the rest of the options in the armature object data is encouraged outside this project. Seeking out tutorials on the subject of armatures will open up many cool options. Unfortunately, the scope of this book only allows a brief skimming of their functionally.

With the armature selected, switch the armature to **Pose Mode** either with the combo box at the bottom of the 3D view or by pressing *Ctrl + Tab*.

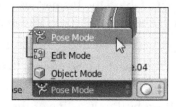

With the armature in pose mode individual bones can be moved or rotated and the mesh will move along with it.

1. Experimentally select (*right-click*) the bone on upper arm.
2. Begin the **Move** (*G*) operation.
3. Move the mouse.

Notice how the bones of the rest of the arm and hand move with it, and the mesh moves with them all too… albeit a little poorly.

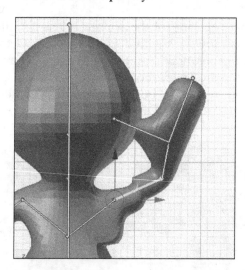

- In the menu at the bottom of the 3D View panel select **Pose | Clear Transform | All** or press *Alt + G, Alt + R, Alt + S* to reset the bone's location.

The reason the mesh didn't move very well is because the effect of the bone on the mesh wasn't very well defined by the skin modifier. It did the best it could, but sometimes it needs a little help. To see the effect that each of the bones has, select (*right-click*) the **BearSkin** and switch to **Weight Paint** mode either using the same combo box at the bottom of the 3D View panel or by pressing *Ctrl + Tab*.

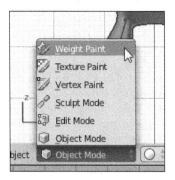

In **Weight Paint** mode the area of effect of each bone can be observed and edited. The red areas are the places that have a 1-to-1 relationship with the bone, matching location and rotation. The rainbow of colors from yellow to green to blue are areas that have less of a relationship with that bone allowing for smooth transitions.

For now just select the various bones to explore how the skin modifier set the bones area of effect. Actually fixing these weights will come later after some more detail is added to the mesh.

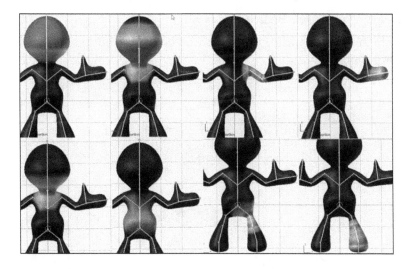

Note that in **Weight Paint** mode selecting the bones does not deselect the mesh. This is a useful function when editing the weights.

Drawing the details

Blender's sculpting tools allow for organic modeling of objects with simple motions. As of version 2.66 those tools were updated with **Dynamic Topology**, making them nearly as good as commercial sculpting tools. In the past sculpt mode worked best with very high resolution models, but now the resolution can be automatically and dynamically changed in local areas keeping the rest of the mesh a more manageable resolution.

Slow down with this part. The tools and techniques taught in this section are very much like drawing, only in 3D. Some users may find this more suited to their tastes. While there may not be as many steps shown expect to take a little extra time doing each step until it looks right. Just like with pen and paper good drawing takes time. Slow down, take it easy, and enjoy the process.

Another tip for drawing is zooming out frequently. Staying zoomed in all the time it becomes difficult to see the whole for the details sometimes. It's best to zoom in, do some edits, zoom out, and look at how that edits looks as part of the whole, and repeat.

As mentioned previously, this work can only provide a high level look at these tools. Exploring the sculpt tools is also encouraged.

1. In the outline view select the eye icon on the **Armature** line to hide the armature.

2. Select (*right-click*) the **BearSkin**.

3. Switch to **Sculpt Mode** with the combo box at the bottom of the 3D View panel.

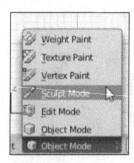

In sculpt mode new tools appear in the left side bar. One of these, dynamic topology, is new to Blender, but allows it to adjust the mesh on the fly to put the detail where it's needed to make good sculpts.

1. Scroll down in the left side bar and find the **Topology** options.

2. Click on the **Enable Dynamic** button.

3. Click on the option that comes up in the pop up menu.

4. Turn the **Detail Size** down to 10.

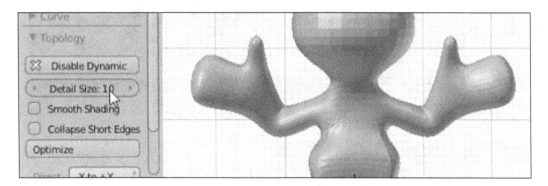

5. Scroll down some more to the **Symmetry** options.

6. Open the options and click on the **X** button under **Mirror**.

Now any edits made to the left side of the model will be automatically reflected on the right saving work and keeping the model symmetrical.

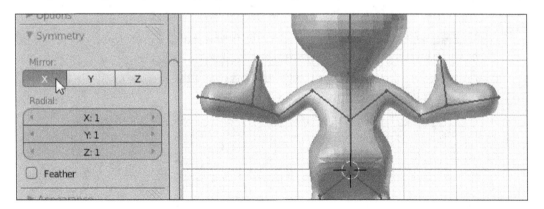

1. Scroll back up in the side bar. Click on the large picture to open up and see all the brushes available.

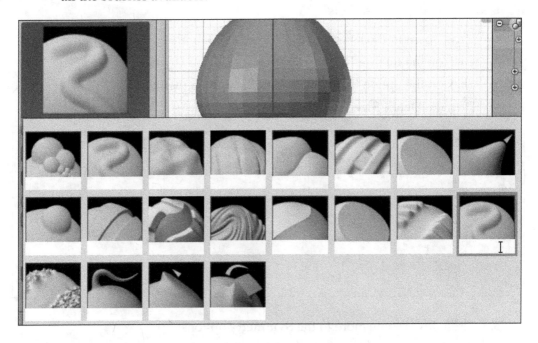

2. Select the sculpt draw brush by clicking on the second image of an s-curve rising out of the mesh or by pressing the *D* key.

This brush is the one most like drawing. It can either build up or cut into the mesh depending if the **Add** or **Subtract** option is selected.

1. To build up the muzzle of the teddy, with the sculpt draw brush (*D*) selected in front view zoom in to the face.

2. Press *F* to adjust the size of the brush and move the mouse until the brush is big enough to draw a muzzle quickly.

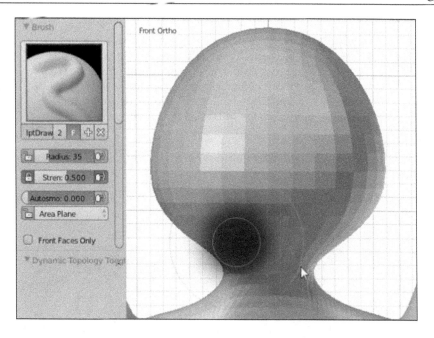

3. *Left-click* and draw smooth circles at the bottom of the head up build up the area. Rotating the view makes the effect more apparent.

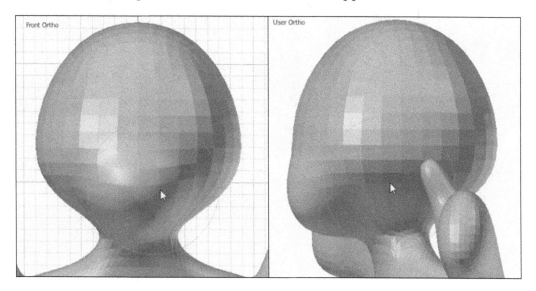

Dynamic topology has been changing the mesh to make the sculpted elements as smooth as possible. Switching to wireframe the effect becomes apparent. Anywhere that's been sculpted has more points. Dynamic topology is closely related to the view. Zooming in more will add more points than drawing more zoomed out. Keep this in mind while working.

4. Rotate the view, choosing an angle where the cheek is visible but not obstructed by the hand.

5. Adjust the brush radius (*F*) a little smaller and sculpt (*left-click*) to build up the cheeks.

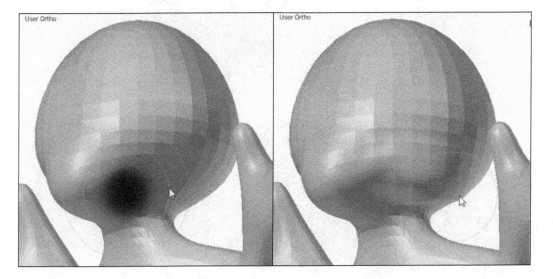

6. Zoom in on the eye area.

7. Adjust the brush size (*F*) and click on the **Subtract** button.

8. Sculpt a recess for the eye.

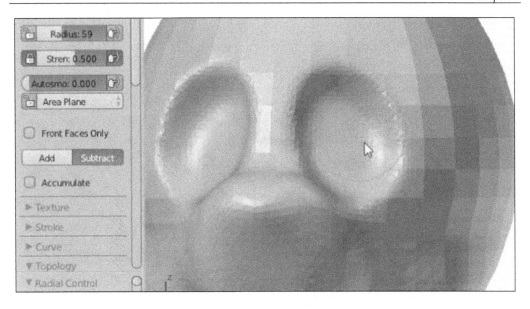

9. Either click on the brush icon and choose the **Inflate** brush, the one that looks like a protruding bubble, or press the *I* key.

10. *Left-click* and gently reinflate the eye recess to make an eye.

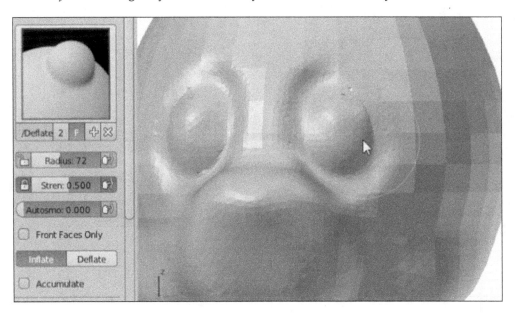

11. Switch to the **SculptDraw** brush (*D*).

12. Zoom in on the eyes and adjust the brush size (*F*) to a smallish point.

13. With the **SculptDraw** brush in **Subtract** mode repeatedly click in the eye to dig a pupil into the eye.

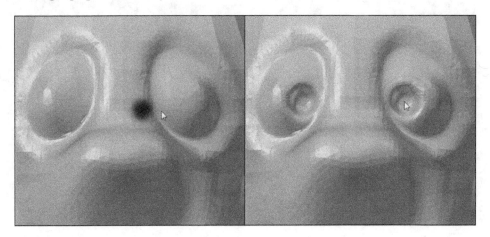

14. Switch to **Add** mode.

15. Draw a bear nose.

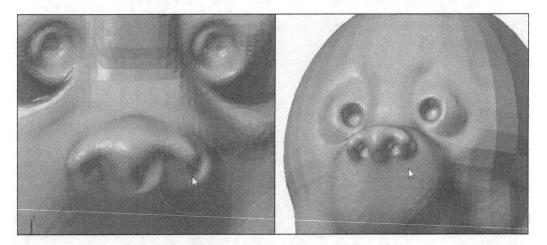

16. With the brush in **Subtract** mode zoom into the mouth area and draw a mouth.

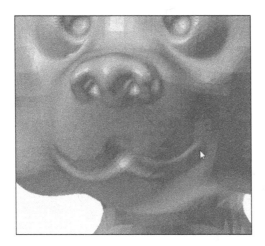

Finally, the bear needs some ears. For this one of the most exciting of Blender's sculpt tools will be used, the **Snake Hook**. The **Snake Hook** must be chosen from the brushes menu by clicking on the current brush and choosing the brush that looks like a snake being pulled out of the mesh.

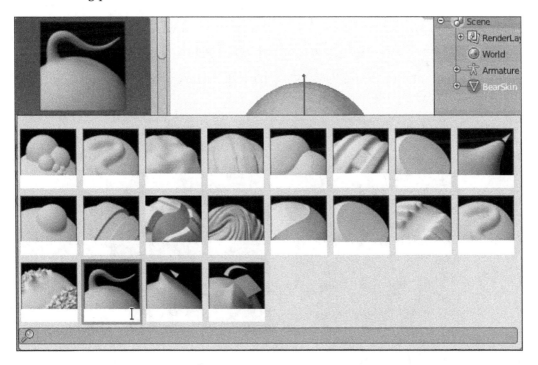

This brush is exciting because with Dynamic Topology it generates whole new areas that could be used to build limbs or appendages. For now only ears are necessary, though.

1. Switch to the **Front** view (*Numpad 1*).

2. Adjust the brush radius very large.

3. With the **Snake Hook** brush, pull a little bit of the side of the head to make bear ears.

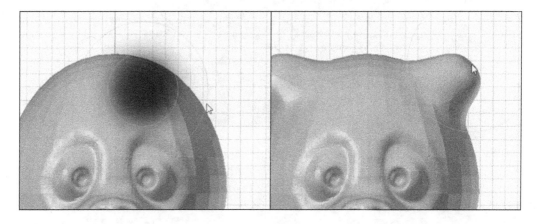

4. Use the **SculptDraw** brush (*D*) with a smaller radius (*F*) in **Subtract** mode.

5. Put a dent in the ear to give it a more bear ear-like shape.

Keep drawing until the desired shape is achieved. Zoom out and look at it from many angles while working.

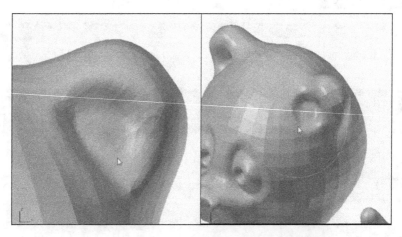

With that the bear is as complete as it's going to be in this work, however there is certainly more that could be done if desired.

Simplifying the model

One of the problems with Dynamic Topology is that meshes can quickly gain a lot of polygons making a more complex shape. With high polygon models even powerful computers can slow down. Fortunately, it is easy to reduce the number of polygons in a model often without making any visible difference in the model, while reducing how hard the computer has to work.

- In the modifier tab on the **Properties** panel on the right press the **Add Modifier** button and select the **Decimate** modifier.

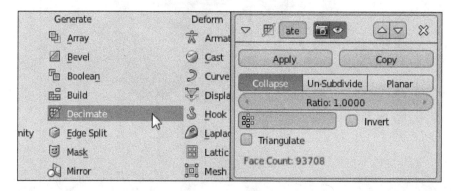

Decimate works by entering a ratio between 0 and 1 and reducing the polygons in the model until they match the ratio. So to reduce the polygons to 1/10 of their original count a ratio of 0.1 would be entered. Decimate attempts to remove polygons that won't be noticed, but if the count is reduced too much it becomes very apparent. The following illustration is the teddy bear model at about 100 polygons, 200 polygons, 500 polygons, 900, and 9000, and 90,000 polygons:

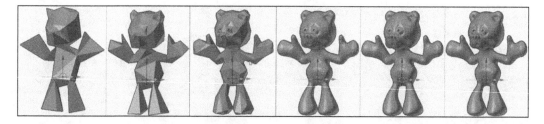

There comes a point of diminishing returns where more polygons have little noticeable effect except in the performance of the computer. With a little trial and effort, turning the modifier off and on with the eye icon to observe the differences the best ratio can be discovered.

In this case that ratio was about `0.50`, but individual experience may vary.

- Set the ratio and **Apply** the **Decimate** modifier

Fixing the armature weights

Sculpting the model added geometry. Decimate took some away. Even if the weights attached to the armature were correct to start with it's very unlikely that they would still be. In order to make the armature useful again the weights need to be repainted.

1. Switch the **BearSkin** to **Weight Paint** mode (*Ctrl + Tab*).
2. In the Outliner panel unhide the **Armature**.
3. Select the **Armature** in the **Outliner** panel and insure that it is in **Pose Mode**.
4. Insure that the **Pose Position** button is selected in the armature object data tab.

With the **BearSkin** in **Weight Paint** mode and the Armature in **Pose mode/Pose Position**, it is time to fix the weights.

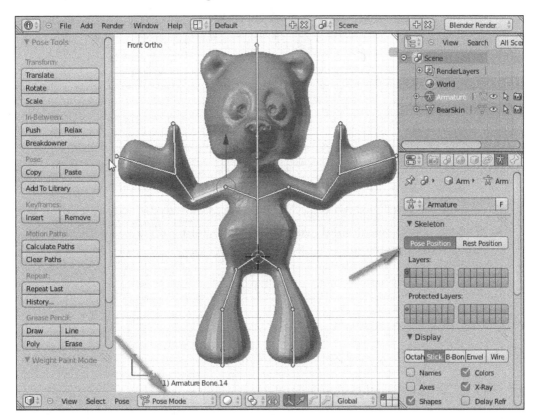

Similar to sculpt mode, the weight paint mode has a set of brush tools, but really there is only one brush necessary. The draw brush has a blend option that duplicates all the other brushes and is considerably easier to find the tool desired.

1. Select (*right-click*) the **BearSkin**.

2. Verify the **Weight** and **Strength** are 1.

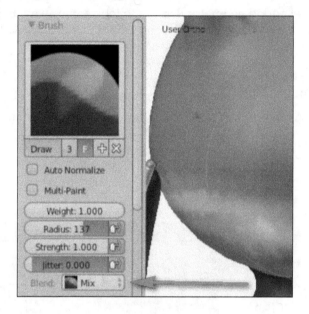

The procedure is as follows:

1. Select a bone. In **Weight Paint** mode selecting a bone will not deselect the mesh.

2. With the **Draw** brush select the **Add** option in the blend combo box.

3. Adjust the brush size with the *F* key (same as in sculpt mode).

4. *Left-click* on the mesh until everywhere that should move when the selected bone moves is red, adjusting the view as necessary.

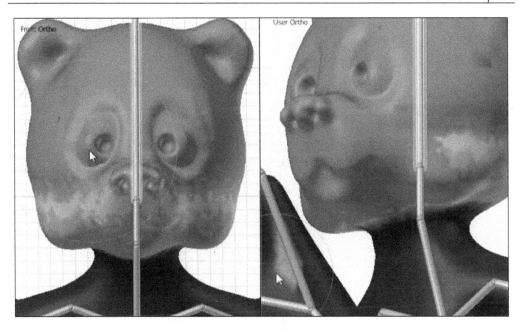

5. **Move** (*G*) the bone and adjust the view to look around.

6. This will likely reveal some part that was missed or some part that was accidently colored.

7. Switch **Blend Mode** to **Subtract** and paint (*left-click*) the part that was accidently painted. This can be done while the bone is posed, and will be instantly applied.

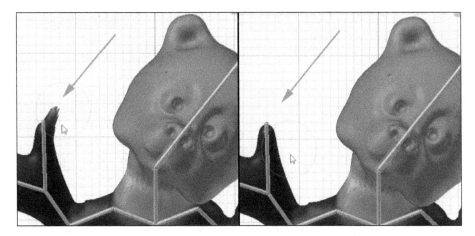

8. When only the right parts are painted red, return the bone to its original position (*Alt + R*).

9. Switch the **Blend Mode** to **Blur**.

10. Soften around the edges of the painted area to extend the gradient. This will insure the movements are slightly more natural.

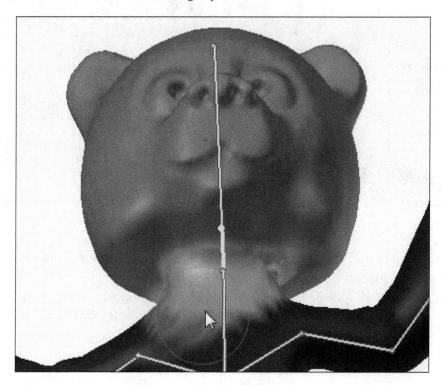

11. Select another bone and repeat the process until all the bones have been corrected.

This is a process that can take a lot of time to do right. Keep at it until the motion is sufficiently smooth. Perfection is not necessarily the goal, but complete coverage and adequacy is.

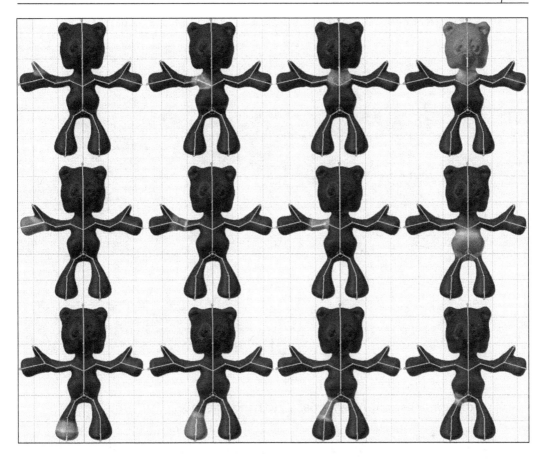

Posing the bear

This bear is a great looking model, but it's not printable the way it is. Keeping in mind the overhang and bridging rules from *Chapter 1, Design Tools and Basics*, the bear needs to be put in a pose that will print. The legs are a clear problem so the bear will be made to sit on a box.

Use the combo box at the bottom of the **3D View** panel to switch from **BearSkin** to **Object Mode**.

1. **Add** (*Shift + A*) a cube to the scene.
2. **Scale** (*S*) and position (*G*) it so it is under the bear's hips.
3. Select (*right-click*) the hip bone in the armature.

4. **Rotate** (*R*) the hip bone around the x axis (*X*) to make the bear sit.

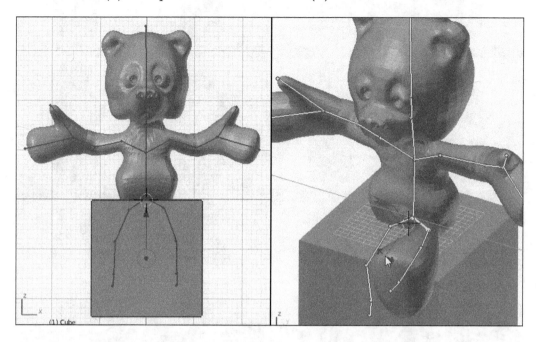

5. Select (*right-click*) the calf bone.
6. **Rotate** it (*R*) around the x axis (*X*) to bend the knee.
7. Repeat with the other leg.
8. Select (*right-click*) and **Rotate** (*R*) the cube around the z axis (*Z*).
9. Position (*G*) the cube in the xy plane (*Shift* + *Z*) to put it properly under the bear.

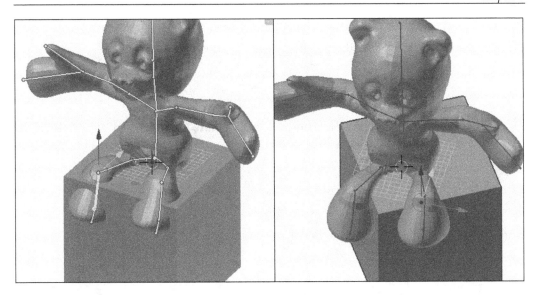

10. Select (*right-click*) the bone through the body.

11. **Rotate** (*R*) it slightly around the x axis (*X*) to get the bear to lean back a little.

12. **Rotate** (*R*) it slightly around the y axis (*Y*) to give the bear some lean.

13. Select (*right-click*) the bear's left upper arm bone.

14. **Rotate** (*R*) it around the y axis (*Y*) to bring the arm down to the bear's side.

15. Select (*right-click*) the left lower arm bone and **Rotate** (*R*) it around the x axis (*X*) to bring the hand to the top of the cube.

16. Select (*right-click*) the left hand bone and **Rotate** (*R*) it around the x axis (*X*) to bring it level to the top of the cube.

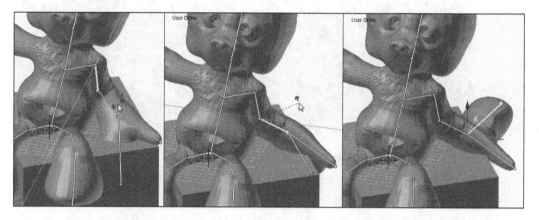

17. Resize (*S*) the cube along the xy plane (*Shift + Z*) and move (*G*) it until the bears hand rests nicely on it.

In this position most of the overhang issues of that hand should be fixed.

1. Select (*right-click*) the right shoulder.

2. **Rotate** (*R*) it to bring the arm up.

3. Select (*right-click*) and **Rotate** (*R*) the forearm around the y axis (*Y*) to position the other arm in a waving position. Don't worry if the arm is intersecting the head slightly as long as it looks like it will print well.

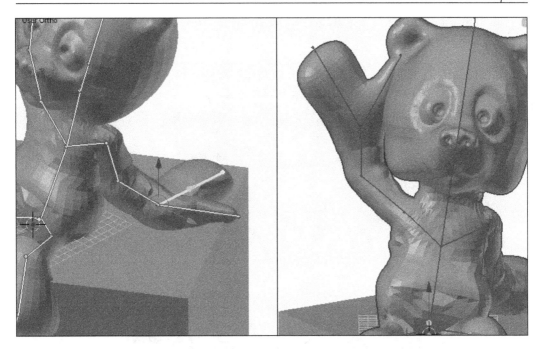

 Technically sticking part of the model into itself makes what is called a self-intersecting mesh and for 3D printing that is generally regarded as bad. However, when printing most slicers will detect and handle the self-intersection gracefully and the model will still print. However, the model could be cleaned up with a mesh cleaning utility. One of the best is http://cloud. netfabb.com. Simply upload the model to them and in a little while download it back with all the bad geometry and self interstations cleaned up ready to print.

4. Select (*right-click*) the neck bone.
5. Press the *R* key twice to enter a special rotation mode called trackball rotation.

6. Tilt the head back slightly to reduce overhang on the chin and position it to look up.

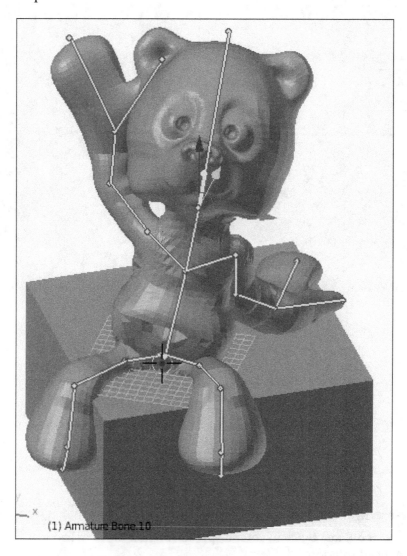

(1) Armature Bone.10

It is perfectly acceptable to go back into weight paint and try to clean up some of the weights to make the modified geometry a little cleaner at this point. Chances are blur is the only brush necessary. Only fix the bigger problem areas. For the smaller problems there will be some clean up in the next step.

Inspecting before print

1. Select (*right-click*) the cube.

2. Duplicate (*Shift + D*) it.

3. Hide all but the duplicated cube (*Shift + H*).

4. **Boolean Union** it with the **BearSkin**.

5. Name this cube `Finalfigure`.

Inspect the `FinalFigure` from all angles, particularly the side views, looking for unprintable areas that violate the 45 degree rule badly. In this case there is at least one area that stands out. The back of the head juts out like a shelf. That's going to cause problems. The tools to fix this part are already known.

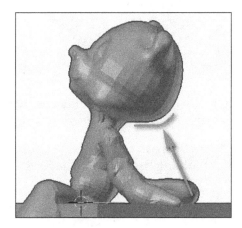

6. Before fixing this **Apply** the Boolean modifier.

7. Enter **Sculpt Mode** using the combo box at the bottom of the 3D View panel.

8. Use a wide brush (*F*) and a combination of the **SculptDraw** brush on subtract and smooth brush to reduce the back of the head until it is a better angle for printing.

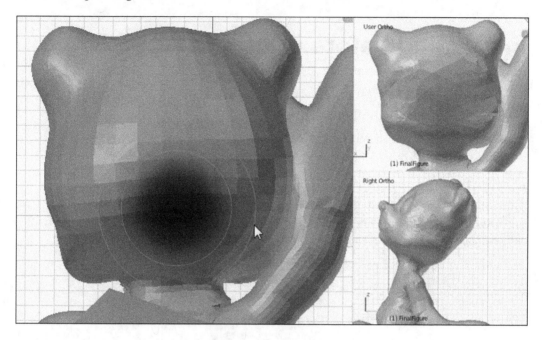

Quad View (*Ctrl + Alt + Q*) may help here to quickly ensure the changes made here are as desired. The feet may also need to be sculpted for print.

While in sculpt mode the **Smooth** brush can be used to fix any other problem areas that result from the armature folding badly. Make the final model as good looking as possible.

When all sculpting edits are complete, switch back to **Object Mode** in the combo box at the bottom of the 3D View panel and **File | Export | Stl (.stl)** the **FinalFigure** for printing. Name the exported file `Teddy Bear.Stl`.

Extra credit

These tools have application beyond making Teddy Bear figurines. The technique in this chapter can be extended to custom game mini-figurines. Also the skin tool can be used alone to make a wireframe model suitable for miniatures like a trellis bridge for a train set or a half constructed building or a bird cage. Many artistic projects have been made using the skin tool over a mathematically interesting mesh that can be used as conversation pieces or plant holders.

Summary

There's more than one way to create a 3D model. The different tools available allow for all kinds of modeling, not just blocky technical parts but clever organic objects that are interesting to look at as well:

- Using the skin modifier to create a basic body shape
- Using the skin modifier to create an armature
- Painting weights and posing the model with an armature
- Using the sculpt tool to make a model with tools similar to drawing
- Using the decimate modifier to reduce the polygon count of a model

At this point all the most common and powerful techniques for creating models from scratch in Blender have been introduced. The next chapter will focus on how Blender can be used to clean up an existing model to make it ready for print.

8

Repairing Bad Models

There are many sources for 3D models on the internet, but not all of them were made for 3D printing. Even if their tolerances are good for 3D printing, the meshes may have holes, flipped normal, or just general bad geometry that will need to be fixed before it can be printed. Being able to find and fix bad geometry will ensure good prints.

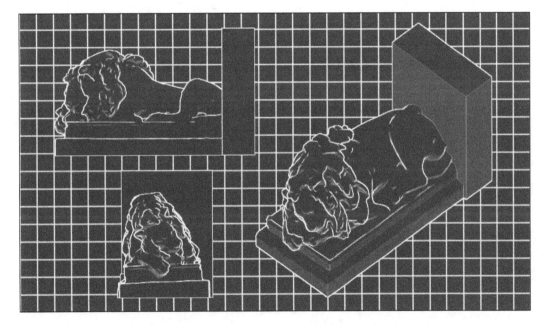

One of the coolest sources for 3D models is from captures of real life objects. There are programs and even online services that can convert a series of photographs into a 3D model. However, it is highly unlikely that those scans will be comprehensive. Sometimes certain angles are difficult to capture and usually the subject being captured is on a surface that cannot be photographed through, such as the ground. Consequently, there will be holes in the scan that will need to be fixed.

In this chapter, a model captured from photographic data and processed by a desktop application will be cleaned up for print using some new techniques and tools. These techniques can be applied to more than just scanned images. Sometimes even models made by hand suffer somewhat from bad geometry, and knowing these techniques will help.

Downloading a 3D scanned file

Go to `http://www.thingiverse.com/thing:90754` and click on the **Download This Thing!** button. Download the `CH8_LionCapture.obj` file and choose to save it in a new folder in the `MakerbotBlueprints` directory named as `Ch`. In Windows this is accomplished, same as in previous chapters, by right-clicking on the link and choosing **Save link as...**, navigating to **Libraries | Documents | MakerbotBlueprints**, clicking on the **New folder** button, and naming the new folder `Ch 8 Scan Repair`. Then double-clicking on the new directory and saving the file.

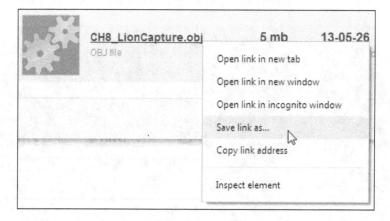

Trimming the fat

After the file is saved, open Blender, clear the scene, and save the scene in `Documents\MakerbotBlueprints\Ch 8 Scan Repair` and name the file `Lion Cleanup.blend`:

Import the file to be cleaned up by clicking on the menu options **File | Import | Wavefront (.obj)**:

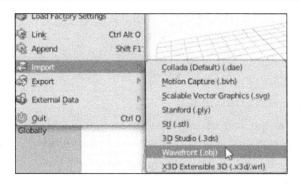

Double-click on the CH8_LionCapture.obj file. The file is very large and may take even a high-end machine a few seconds to process. When imported, it may be difficult to find the subject of the scan in all the extra data. The subject will be near the middle. Zoom in and adjust the view until the lion in the middle is framed in the view.

When zoomed in, it becomes clear that the lion is at a strange angle to the work plane.

3D scans often capture more than the focus of the scan. All that extra, loose geometry needs to be separated and eliminated, which will make the model much easier to manage. To do this the Separate command will be used:

1. Select (*right-click*) the scanned object.

2. Press *Tab* to enter **Edit Mode**.

3. In the menu at the bottom of the **3D View** panel, choose **Mesh | Vertices | Separate** or press *P* to select the Separate command.

4. In the pop-up menu, click on **By loose parts**.

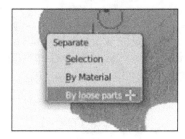

5. Press *Tab* to exit **Edit Mode**.

6. Select (*right-click*) the lion object in the center.

7. Press *Ctrl + I* to invert the selection.

8. **Delete** (*X*) all the selected objects. Only the lion will remain.

Trimming the fat in this way will also speed up Blender since there are much fewer points to process.

Orienting the scan

The lion is still oriented wrong. By changing the camera view, it becomes apparent that this model is oriented to the y axis up and down. This is simple enough to fix by rotating (*R*) the object -90 degrees around the x axis (*X*).

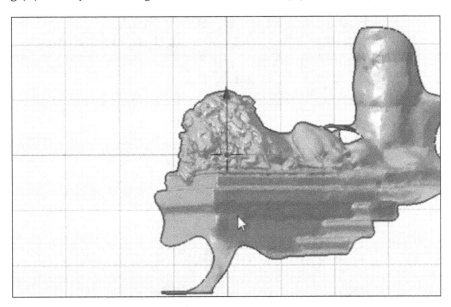

Trimming more fat

There are extra parts of the model that won't be needed. The easiest thing to do will probably be to just trim away the excess.

1. Enter **Edit Mode** (*Tab*).

2. From the **Front** (*Numpad 1*) or **Right** view (*Numpad 3*), the **Ortho** view (*Numpad 5*) in the **Wireframe** view (*Z*), **Border Select** (*B*) all the points under the pedestal the lion is sitting on.

3. Take special care to get all the points under the line. Zoom out if necessary.

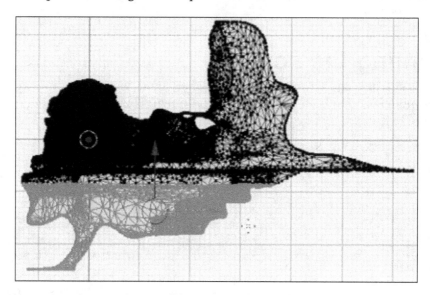

4. **Delete** (*X*) the selected **Vertices**.

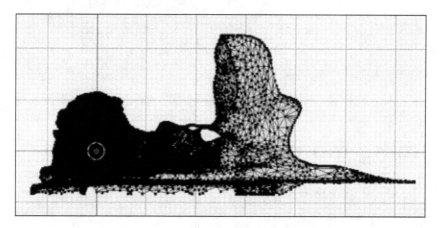

Cleaning up the back will be much easier if the model is lined up with the gridline of the XY plane, so line the model up using **Rotate** (*R*).

5. Exit **Edit Mode** (*Tab*).

6. View the model from the **Top** (*Numpad 7*).

7. **Rotate** (*R*) the model until the pedestal lines up better with the gridlines.

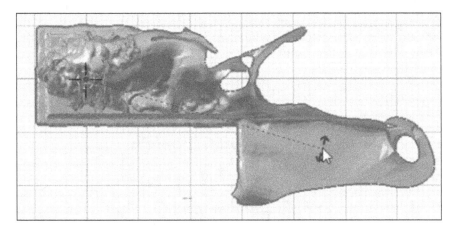

8. Now to do a little more trimming, switch back to **Edit Mode** (*Tab*).

9. In the **Wireframe** view (*Z*), **Border Select** (*B*) the extra geometry behind the lion.

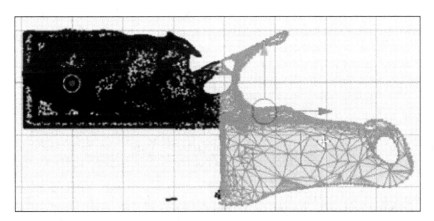

10. **Delete** (*X*) those **Vertices**.

If there are any loose vertices left, clean them up by selecting (*right-click*) and deleting (*X*) them.

Making a flat base

While inspecting the lion it becomes clear that all it is right now is a thin shell with huge gaps in it which is not very printable. What needs to be done is to clean up the mesh so it is closed, manifold, and has a flat bottom suitable for printing.

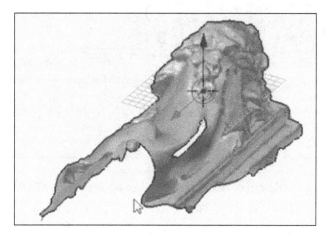

A solid, printable mesh is closed, or watertight, and manifold.

Closed or watertight is an easy enough concept to get. In cartoons, the characters can blow bubbles of any size or shape they want, but if that bubble gets a single hole it pops. In the same way, the mesh should be one continuous surface no matter the twists and turns it takes.

Manifold is a mathematical term that in general terms can be confusing, but in specific turns for our purposes means that every edge sits between no more than two faces and no face intersects any other face. (If it's watertight as well, each edge will border exactly two faces.) If an edge is bordering three faces, then there is an unnecessary face in the model. Non-manifold meshes can make it difficult for the slicer to tell what is supposed to be "inside" or "outside" the model.

Proper normal orientation is also important, but if the mesh is watertight then manifold fixing the normals is no problem at all.

Blender has a special function specifically to select non-manifold portions of the mesh.

1. Enter **Edit Mode** (*Tab*).

2. Select non-manifold points by choosing **Select | Non Manifold** in the menu at the bottom of the **3D View** panel or press *Shift + Ctrl + Alt + M*. The following selected points are what Blender identified as being non-manifold and therefore need to be fixed:

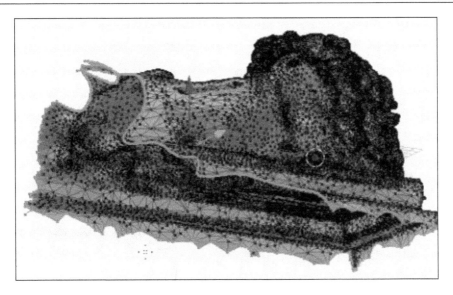

At the moment there aren't that many non-manifold parts. Most of them are simply the edge loops that need to be closed.

3. With the non-manifold points selected, use the **Circle Select** (*C*) and deselect (*middle-click*) all the points, not on what could be considered the bottom.

4. When done the only points that should be selected are the ones that will be extruded to form the bottom.

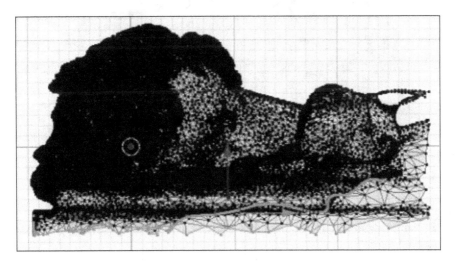

5. **Extrude** (*E*) the selected points down along the z axis (*Z*).

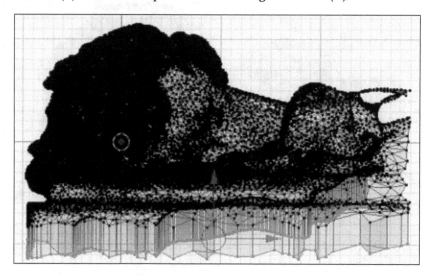

6. Flatten the bottom by scaling (*S*) the extruded points along the z axis (*Z*) by zero (0). Now the bottom is flat, but it is still not filled in.

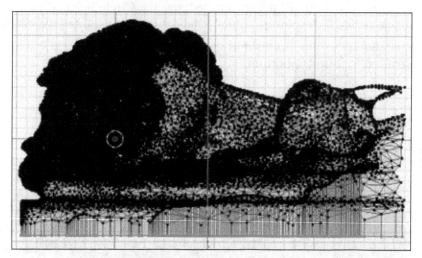

7. **Extrude** (*E*) the selected points but do not touch the mouse.

8. Press *Enter* so a new set of points is created from the selection in the same location.

9. In the menu at the bottom of the **3D View** panel, navigate to **Mesh | Vertices | Merge** or press *Alt + M* to merge and click on **At Center** from the menu that pops up.

All the selected points will be brought together and turned into one point at the calculated center.

Inspect the new bottom and make sure the connecting lines do not cross the edge of the bottom to get to the merged point. Depending on a number of factors, it is possible that this has occurred as indicated in the following screenshot:

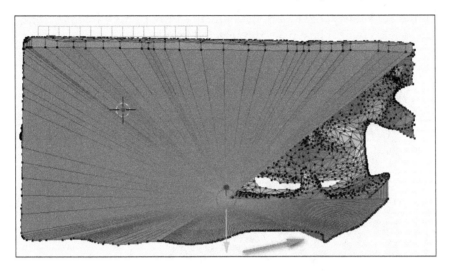

If this happens, the simple fix is to **Move** (G) the center point along the XY plane (*Shift* + *Z*) until it is at a place, where all the connecting lines can reach their respective edge point without crossing the edge anywhere else.

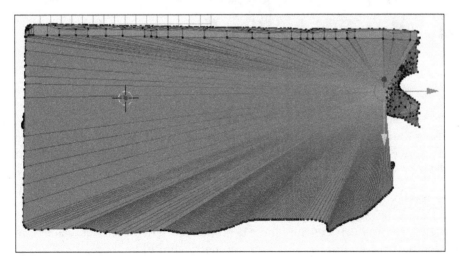

There is still a wedge that remains unfilled on the bottom. To fill this wedge, perform the following steps:

1. Hold the *Shift* key and select (*right-click*) the vertices of either edge of the wedge to be filled.

2. Three points will be selected.

3. In the **3D View** panel menu, navigate to **Mesh | Faces | Make Edge/Face** or press *F* to create a face from the selected points.

Detail work on the back

The back of the lion will be extruded and sealed in much the same way, but before it can be there is some extra geometry from the scan that needs to be identified and cleaned up first.

Zoom in on the posterior of the lion and locate the plate of points that are just dangling there, barely attached to anything shown in the following screenshot. Because more or less vertices may have been deleted when the part was trimmed what is left may differ, but the spurious geometry should be there.

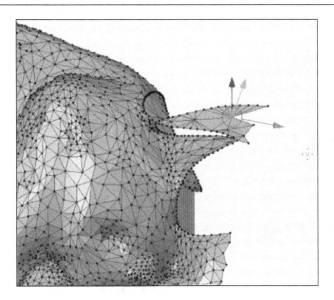

Select these points using any desired method:

1. **Border Select** (*B*), **Circle Select** (*C*), or select (*right-click*) the points in the middle and expand the selection (*Ctrl + Numpad +*) until the desired points are selected.

2. Deselect (*middle-click*) the point at the very tip (indicated in the following screenshot) since that point is shared with the main body of the lion and should probably be left.

3. Then **Delete** (*X*) the selected points.

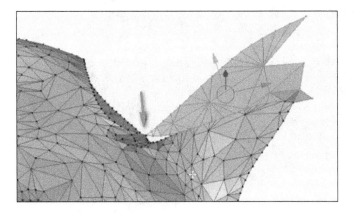

4. When those points are deleted, select the spur sticking out of the body next to them. It is harder to determine where to stop, so some judgment must be applied.

5. Again, select in any desired way and **Delete** (*X*) the spurious vertices.

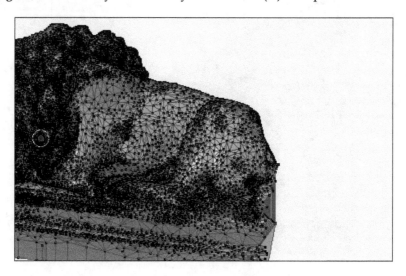

6. Then, same as with the base, use the **Non Manifold** command (*Shift + Ctrl + Alt + M*) to identify the edges that need to be sealed off. **Circle Select** (*C*) and deselect (*middle-click*) (probably best in the **Wireframe** view) all, but the back edge. This is complicated because the editing done on the bottom has likely introduced new non-manifold points. Don't worry about those points for now and just deselect all but the back edge.

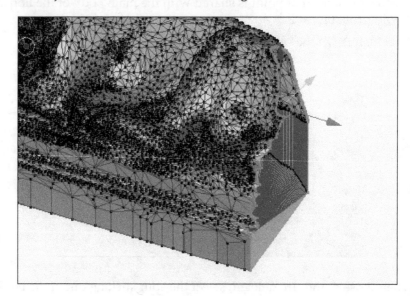

7. Same as before, **Extrude** (*E*) the selected points.

8. **Scale** (*S*) them, this time along the x axis (*X*) by 0 units to flatten them.

9. **Extrude** (*E*) again, this time without moving them before pressing *Enter*.

10. **Merge** (*Alt + M*) **At Center** the points to seal the back of the lion.

11. **Move** (*G*) the merged point along the YZ plane (*Shift + X*), so that all the edge lines reach the edge without crossing the edge anywhere else.

This merge point may be more difficult to align to make all points happy than the last time on the bottom. For now, good may have to be good enough.

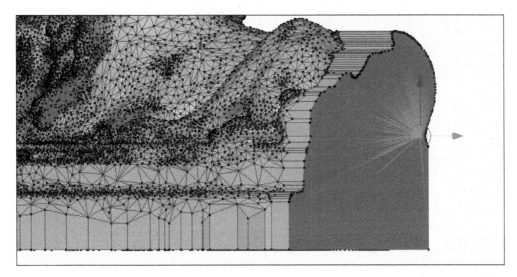

Now, there is only one hole left to close off, the one on the back of the lion.

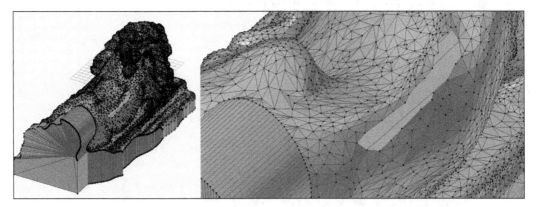

To quickly select all the points on the edge of the hole, use the `Loop Select` command.

1. Press and hold the *Alt* key.

2. Select (*right-click*) any of the lines on the hole, not any of the vertices. Blender will attempt to find the loop that the selected line is a part of.

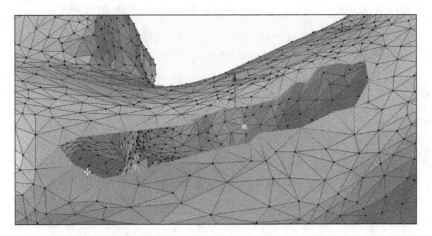

 Loop select doesn't always work when the geometry is complicated, which is why it wasn't demonstrated earlier in this project. But in this case loop select works just fine.

3. Click on **Make Edge/Face** (*F*) to make a face from the selected points.

Since this hole is simpler and less complicated than the previous ones, Blender will fill in the face correctly this time.

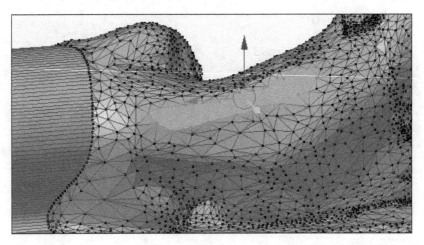

Cleaning up bad geometry

Now, the model should be watertight, but it is hardly clean and non-manifold. Now is the time to locate, identify, and fix the non-manifold points:

1. Clear the selected points (*A*).

2. Select all non-manifold points (*Shift + Ctrl + Alt + M*).

Unfortunately in sealing the edges, non-manifold geometry was accidently introduced into the model. All of the currently selected points represent a problem in the model that should be fixed.

The first step is to remove any points that are occupying the exact same spot:

1. Select all the points (*A*).

2. Click on the **Remove Doubles** button in the left-side bar.

3. When that is done clear the selection (*A*).

4. Select the non-manifold points again (*Shift + Ctrl + Alt + M*).

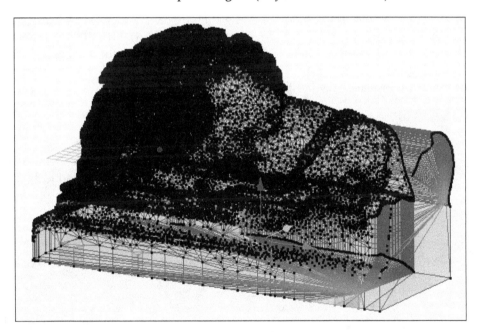

5. Choose one of the selected points and zoom in on it.

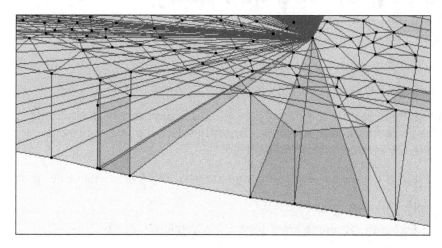

Deleting an extra edge

The previous screenshot is zoomed in on one such place. The darker grey areas indicate that there are multiple surfaces overlapping each other in that area, two faces on top of each other. There is some edge on the border of three faces, which is one of the checks for non-manifold. The best thing to do is identify the extra edge and delete it. In this case, it is fairly simple to guess:

1. Select (*right-click*) the vertex on the left of the dark area.

2. Additionally, select (*Shift + right-click*) the vertex on the right.

3. These two points should not be connected by an edge since they would be going through the point in the middle. However, the red line connects them clearly, so this is the errant edge.

4. **Delete** (*X*) that edge between these two points.

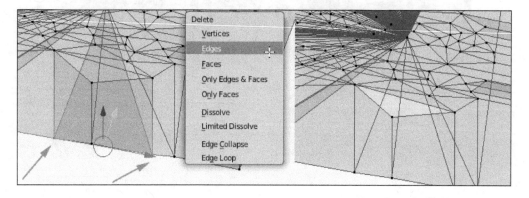

Unfortunately this introduces a new problem. The lighter triangle that appears is a new hole in the geometry. Apparently that edge was a part of a face that wasn't covered by the other points. That will have to be fixed.

5. Loop select (*Alt + right-click* on an edge) the hole.

6. Create a face (*F*) from the selection.

Merging the problem away

Another way to fix geometry problems like this is by using the **Merge** tool. The illustrated section is another area identified as non-manifold.

1. Select one of the involved vertices that is at the bottom corner of the mesh.

2. Select the other five vertices one at a time with the *right-click* and holding the *Shift* key.

3. Then **Merge** (*Alt + M*) and choose **At First**. The position of the first chosen vertex will be where the selected points will be merged to.

The **At First** or **At Last** options do not appear and are not affected by using **Box Select** or **Circle Select**. At least one point must be selected with *right-click* for those options to appear. By using this option, more control can be exerted over where the merged points will meet.

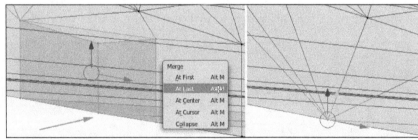

Finding hidden points

Another situation that may cause problems is unnecessary internal geometry. If a certain point is visible in the **Wireframe** mode that cannot be seen in the solid view from any angle, it is safe to assume that the point in question is entirely unnecessary and safe to just delete.

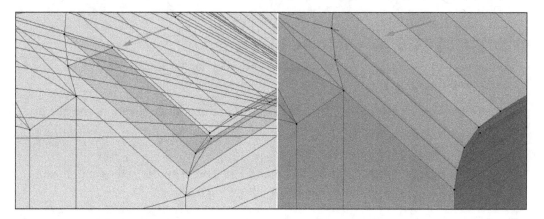

Uncrossing the lines

If the problem is, as illustrated in the following screenshot, the edges that pass outside their geometry such as the merged points on the back, crossing a concave part of the edge loop, the solution is to turn the edges so they connect elsewhere:

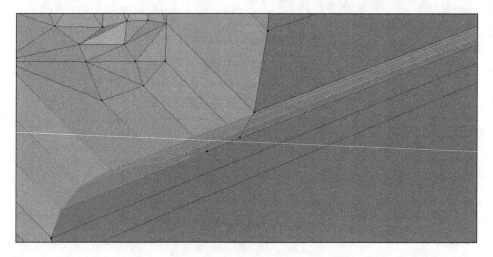

1. Switch to Edge Select mode by clicking on the cube in the bar at the bottom of the **3D View** panel or by pressing *Ctrl + Tab* and then choosing **Edge**.

2. Select an offending edge. It is often best to work from the outside in.

3. Open the **Edge** menu either by choosing **Mesh | Edge** in the bottom menu of the **3D View** panel or by pressing *Ctrl + E*.

4. Select **Rotate Edge CW**.

5. If necessary repeat **Rotate Edge CW** until the edge is no longer crossing the concave space.

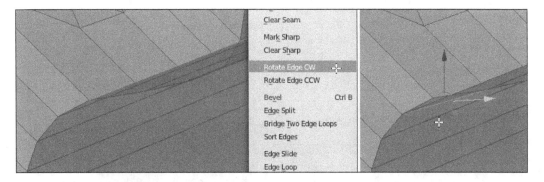

6. Select the next offending edge and repeat the previous steps until all edges are properly aligned.

7. Switch back to **Vertex Select Mode** (*Ctrl + Tab*) if desired.

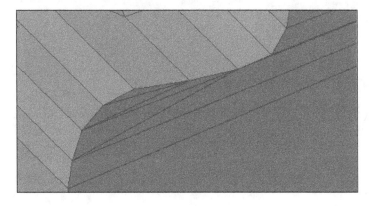

Repeat until clean

It is difficult to identify all the ways that show that geometry can be bad, but these are the most common problems. For knowing these techniques, the steps are simple:

1. Use the **Non Manifold** select (*Shift + Ctrl + Alt + M*) to locate bad geometry.

2. Identify what the problem is.

3. **Delete**, **Merge**, **Move**, or otherwise correct the bad geometry.

4. Repeat until **Non Manifold** selects no points at which the model is manifold.

Final cleanup

The last step when the model is manifold is to verify that the normals are oriented properly. The last time normals were an issue was in *Chapter 3*, *Face Illusion Vase*. There they messed up the way the Skin modifier handled its task. Messed up normals can also affect the printability of models. Fortunately the fix is often simple in Blender:

1. Select all points (*A*).

2. Click on the **Recalculate** button under **Normals** section in the left-side bar of the **3D View** panel.

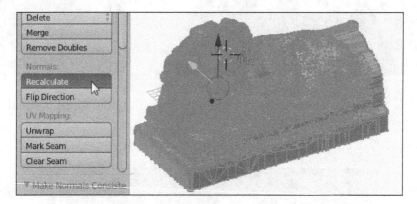

Making it a book end

As-is the back of this model is hardly presentable. One option would be to use the sculpt tools from *Chapter 7*, *Teddy Bear Figurine* to sculpt the missing parts of the lion. An easier option is to throw a cube behind it to hide the blemishes and call this project a book end:

1. Add (*Shift + A*) a **Cube**.

2. **Scale** (*S*) and **Move** (*G*) the cube until it is slightly taller and wider than the statue, and thick enough to hide the back of the lion. Try to be as precise as possible when positioning the bottom of the cube to the bottom of the statue, but don't stress too much about it. The bottom will be re-flattened.

3. Add a **Boolean** modifier to the cube.

4. Union the **Lion_Capture** object to the cube.

5. Apply the modification.

6. Enter **Edit Mode** (*Tab*).

7. In **Right** side (*NumPad 3*) view in the **Wireframe** view (*Z*), **Border Select** (*B*) all the points that make up the bottom of the lion.

8. **Scale** (*S*) along the z axis (*Z*) by 0 units to assure the bottom is perfectly flat.

9. Exit **Edit Mode** (*Tab*).

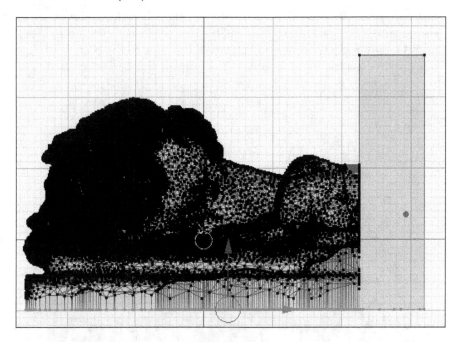

Before exporting the model, **Scale** (*S*) the object to be of desired size. A proper book end should be about 15 cm to 20 cm tall and weighted. To just see the model, something about 5 cm tall should be fine. (Remember that in Blender, cm are the large grid lines, mm are the small ones.) Navigate to **File | Export | Stl (.stl)** and export the model as `Lion Statue.stl` for print.

Summary

When 3D modeling is focused on the render for a video or image, "if it looks good" that's all the criteria the modelers need. However, 3D printing needs to have its geometry well defined, or it can't bring something from the virtual to the real world.

Fixing bad geometry may not be the most entertaining part of 3D modeling for everyone. But being able to identify the unnecessary edges or vertices, create faces to patch holes, and rotate edges when they're in the way, and build a watertight, manifold model are valuable skills. And if you think about it like a puzzle to solve, it can be quite satisfying. Being able to take a model that wasn't made for 3D printing and fixing it, means that entire libraries of ready-made models open up online, so you don't have to make everything you want to print from scratch.

There are tools such as netfabb (`http://netfabb.com`) and meshlab (`http://meshlab.sourceforge.net/`) that can do a lot of these sorts of things semi-automatically. They're not complete solutions for every problem, but often can reduce the amount of work necessary to do mesh correction. But even with these tools models that look good may not always be printing without a little tweaking by hand.

The next chapter will focus less on modeling or model manipulation. Instead, the focus will be on learning how the slicer can be manipulated to change the way a model prints, sort of 3D printing post-processing.

9
Stretchy Bracelet

There is more that can be done to make cool 3D prints than just modeling. Settings on the slicer can create post-processing effects that can alter the final version of a print.

The modeling in this project will be simple and should be nothing new at this point. Instead, the focus will be on the printing process. A bracelet that takes advantage of plastic's natural elasticity is a perfect example for this project.

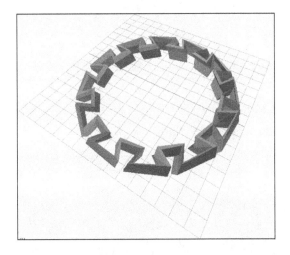

Modeling the bracelet

It should only take a few minutes to model the bracelet, but don't worry about making it hollow. That will be accomplished with the post-processing. Instead, just make sure the top and bottom are flat for this to work. Perform the following steps for modeling the bracelet:

1. As usual begin the new project by clearing the scene and saving it in a new directory under **MakerbotBlueprints**.

2. Name the new directory Ch 9 Stretchy Bracelet.

3. Save the project as **StretchyBracelet.blend**:

4. Add (*Shift + A*) a **Cylinder** to the scene.

5. Change the options of the cylinder to have **24 Vertices**, **Depth** of **6** to **12**, and **Radius** about as big as the widest point of the hand of the person who will be wearing this, usually around the thumb knuckle. This will be about 25 for a child, 35 for an adult female, and 45 to 60 for an adult male.

6. Name the cylinder BraceletShape.

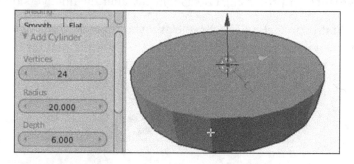

7. Enter the **Edit Mode** (*Tab*).

8. Switch to Face select (*Ctrl + Tab*).

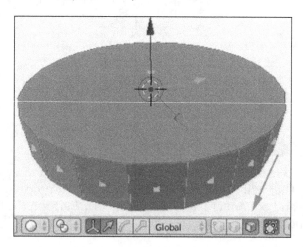

9. (De)select (*A*) all faces.

10. Select (*right-click*) on just one of the faces around the edge.

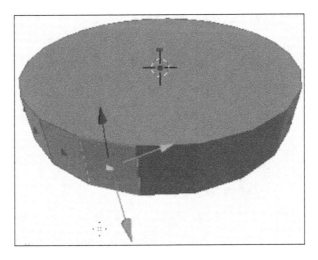

11. **Extrude** (*E*) the face by 4 units.

12. **Scale** (*S*) in all directions but the z axis (*Shift + Z*) by 2 units.

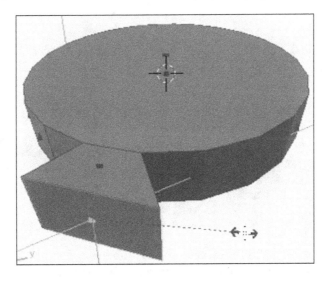

13. Then skip one face around the **BraceletShape** cylinder and select (*right-click*) the next one.

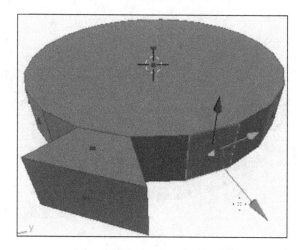

14. Again **Extrude** (*E*) that face by 4 units, and then **Scale** (*S*) in all directions, but the z axis (*Shift + Z*) by 2 units.

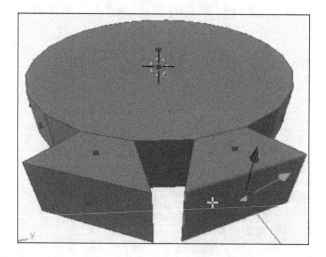

15. Continue around the **BraceletShape** cylinder by selecting (*right-click*) every other face and extruding (*E*) by 4 units, scaling (*S*) in all directions but the z axis (*Shift + Z*) by 2 units.

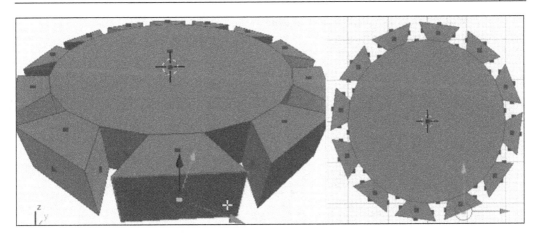

It is important that the top and bottom are perfectly flat. That is why the scale operations were constrained to all but the z axis by pressing *Shift + Z*. If any scale operation were not constrained by *Shift + Z* don't panic. Simply flatten the top and the bottom as was shown in *Chapter 2, Mini Mug*. In **Wireframe** view (*Z*), **Border Select** (*B*) the points on top, **Scale** (*S*) them along the z axis (*Z*) by 0 units, then repeat on the bottom. The side effect of this method is that the height of the bracelet will not be exactly what it was set to at the start, but for something like a bracelet precision is not necessarily important.

Refining the shape

This is technically all that is necessary for this exercise, the outside edge as defined will make a shape that will flex and spring. But it's not as appealing a shape as it could be. With a few simple steps as follows, this boring bracelet can be prettied up somewhat:

1. Return to **Vertex**, select (*Ctrl + Tab*).

2. Set **Wireframe** view (*Z*) and jump to the **Front** (*Numpad 1*), **Side** (*Numpad 3*), and the **Orthographic** (*Numpad 5*) view.

3. **Border Select** (*B*) all the points that make up the top of the **BraceletShape** cylinder.

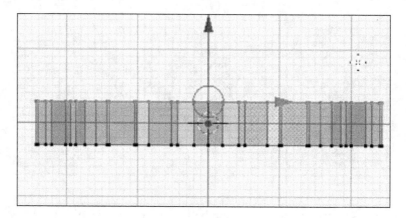

4. **Rotate** (*R*) the selection around the z axis (*Z*) by about 5 degrees.

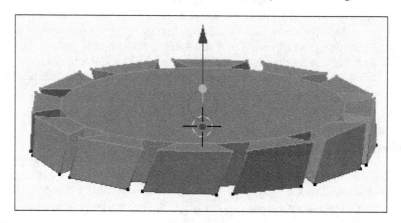

5. **Loop cut** (*Ctrl + R*) around the middle of the **BraceletShape** cylinder.

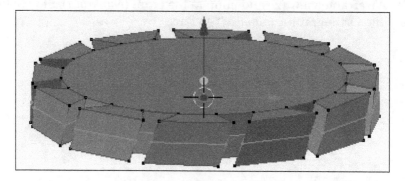

6. **Scale** (*S*) the loop cut so that the bracelet bulges just a little.

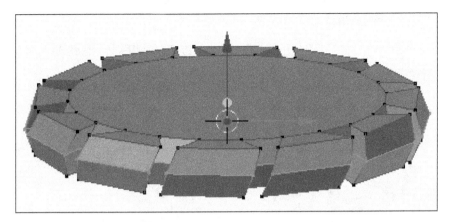

7. Exit the **Edit Mode** (*Tab*).

8. Export the **BraceletShape** cylinder to a **Stl** file (**File | Export | Stl (.stl)**).
   Save the **Stl** file as `Bracelet.stl`.

There, that is a much more appealing shape. Just because this project is academic doesn't mean it has to be boring. Plus this illustrates the fact that as long as the top and bottom are flat, the shape in-between can have all sorts of interesting geometry and this will still work.

Printing the bracelet

With the flat top and bottom, the model may not look much like a bracelet as-is. If printed with normal settings it certainly wouldn't be. However by manipulating the settings on the slicer, the program that prepares models for 3D print, this model can easily be made into a bracelet. How the settings are edited depends on the slicer. There are many slicer programs, just as there are many 3D printers. Makerbot has two official slicers that it recommends, **ReplicatorG** and **Makerware**. To keep the conversation simple only these two will be discussed, but in general the goal is to locate the fill layers and set them to 0 units.

Editing the settings in ReplicatorG

The following steps help to edit the settings in ReplicatorG:

1. In ReplicatorG, first navigate to **File | Open** and locate **Bracelet.stl**.

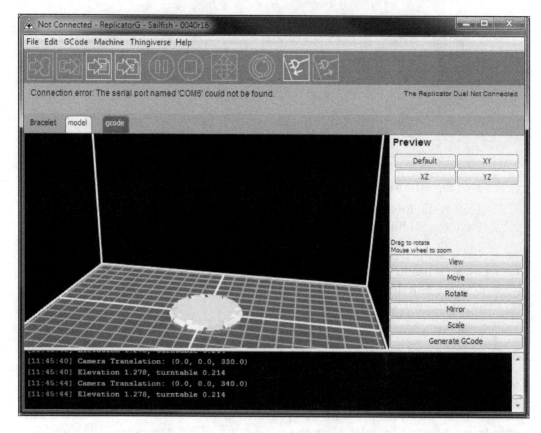

2. In the menu, navigate to **GCode | Edit Slicing Profiles...**.

3. From the **Edit Profiles** menu, choose whatever default profile is normally used and press the **Duplicate** button:

4. Name the new profile NoFill.

5. In the **Skeinforge Settings** menu, locate the tab for **Fill** and click on it.

6. Scroll down and find the **Solid Surface Thickness (layers)** option. Set this option to 0.

7. If the option for **Enable automatic solid surface thickness** is present, uncheck this option as well.

8. From the top menu, navigate to **File | Save and Close**.

9. Press the **Done** button on the **Edit Profiles** menu.

With the new settings profile made, simply prepare the model for printing as usual:

1. Press the slice to the **GCode** button on the top bar.
2. Allow ReplicatorG to move the bottom of the model to the build surface by answering **Yes** to the dialog box that pops up, and save the model.
3. In the **Generate GCode** menu, make sure the **Slicing Profile** field is the new **NoFill** profile just created.
4. Turn off **Use Raft/Support**.
5. Then change the **Object infill (%)** field to 0, and increase the **Number of shells** field to 2 or 3.

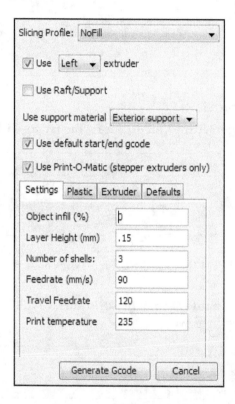

6. Click on **Generate GCode** and print the model either by sending it to an SD card or printing over USB, as usual.

The result will be the model without a top or bottom suitable for a bracelet.

Editing the settings in MakerWare

MakerWare, MakerBot's own slicing program, has many improvements over ReplicatorG in how it slices models and the detail level possible. However, since its settings are so well dialed-in for Makerbot 3D printers they aren't as exposed as they are in ReplicatorG. But they are there and are editable. Perform the following steps to edit the settings in MakerWare:

1. In MakerWare, navigate to **File** | **Open** and locate **Bracelet.stl**.

2. Answer **Yes** to the prompt to put the object on the platform.

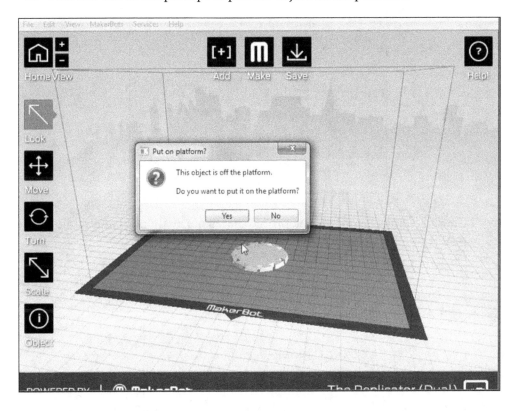

3. Click on the **M** button at the top middle to prepare this model for printing.

4. In the menu that pops up, click on the triangle next to **Advanced** to open-up the advanced options.

5. Then click on the **Create Profile...** button.

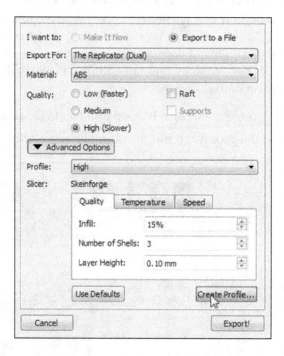

6. Choose from a **Template**, the profile that is normally used and name the new profile NoFill.

7. Click on **Create**.

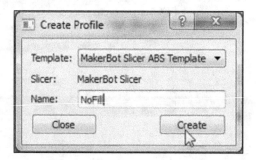

8. Click on the **Edit Profile** button.

9. Most Windows users will see a prompt about opening a **miracle.json** file. Click on **Select a program from a list of installed programs** and then click on **OK**. In the menu that comes up, find and click on **Notepad** and then **OK**.

 Regardless of your system, JSON files are really just text so any text editor will do the job.

10. Search for a line with **infillDensity**.

11. Edit the number next to it to 0.

12. Search for a line with **numberOfShells**.

13. Edit the number next to it to be 3 or 4.

14. Search for a line with **roofThickness** and below that **floorThickness** options.

15. Edit the values next to them to both be 0.

```
miracle.json - Notepad
File  Edit  Format  View  Help
{  "comment"  :  [
  "This is a custom profile for MakerBot Slicer. Editing it will modify your",
  "slice settings. For documentation on MakerBot Slicer parameters, see here:",
  "http://www.makerbot.com/support/makerware/documentation/slicer"
],

  "infillDensity": 0
  "numberOfShells": 4
  "insetDistanceMultiplier": 0.97
  "infillShellSpacingMultiplier" : 0.70
  "roofThickness" : 0
  "floorThickness" : 0
  "layerwidthratio": 1.182
```

Save and close the file and return to MakerWare. Then click on the **Export!** button and either save to an SD card or print over USB, as usual. The result will be the model without a top or bottom suitable for a bracelet.

Summary

Just like being able to modify the settings on a 2D printer allows for things like double-sided prints or printing multiple pages on the same sheet, manipulating slicer settings has advantages as well. In this chapter, a simple model was modified, not by modeling but by changing the print settings to turn a modified cylinder into a bracelet. Alternatively, turning the top and bottom fill layers off, but turning the infill on is an interesting way to make a perfect screen with a controlled mesh.

Some slicers used for other 3D printers don't allow access to settings such as the slicers for Makerbot but most RepRap based printers do. In that case, it would become necessary to remove the top and bottom faces and modify the bracelet to just be a minimally thick wall in Blender. The problem with this is that many slicers, even professional level ones, have a bug that shows up with walls that are of the wrong thickness. A gap appears in the middle of the wall, where the slicer wants to put a shell, but it can't, and just gives up on the area instead of filling it in. Maybe one day this problem will be solved, but until then using the method described in this chapter eliminates worry about wall thickness, since the thickness is a side effect of the process and is therefore perfect every time.

Editing the top or bottom options separately makes printing things such as cups from cylinder shapes possible. Another setting that is worth exploring is the multiply setting that allows making multiple copies of a thing without increasing the time to slice it by much. Becoming familiar with and taking control of other slicer settings can also improve the quality of prints dramatically. Exploration is encouraged; default is not the only option.

The next and final chapter will be explored using the measuring tools and techniques to more accurately produce models that match the needs of the real world.

10
Measuring – Tips and Tricks

When designing things for 3D printing in the virtual space it is sometimes easy to forget the relationship they'll have to the real world when printed. And sometimes it is exactly how they'll interact with the real world that is the point of the design, such as when printing an end cap for a pipe or printing a replacement to a broken part. So, it is often very important that accurate measurements must be made when planning and making blueprints.

Part of this blueprint requires a physical object to follow along. Later, when the real world part is complete, a download will be provided so the reader can follow along as in other chapters.

All of the necessary modeling techniques have been covered in previous chapters. This chapter will focus on measuring real life things, and will teach an interesting technique for transferring those measurements to the virtual space easily.

Using a caliper

The most common technique for taking accurate measurements is the use of a tool called a caliper, a must have for anyone who models for 3D printing. Calipers measure distance with a high degree of precision and can measure in three different ways; the outside diameter of an object with the outside jaws, the internal diameter with the inside jaws, or the depth with the depth probe at the far end. The easiest kind of caliper to use is the digital type.

Simply turn on the digital caliper, tare or zero the reading while closed, then open the jaws, put the object to be measured between them, clamp it down, and take the reading. It's fast and relatively accurate without much effort. Most models even have a port that can transfer the measurements directly to the computer. But digital calipers have the disadvantage of being more expensive and relying on batteries which when run out eliminate the ability to measure accurately.

If budget is a concern then perhaps a Vernier caliper is preferable. These calipers operate purely mechanically, but have a clever trick that allows them to be just as accurate if read properly.

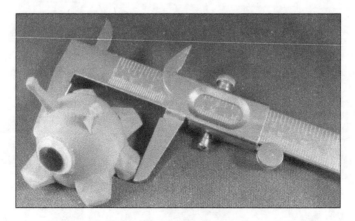

Again, open the jaws that take the measurements and tighten it to the object to be measured. Then take a close look at the little window.

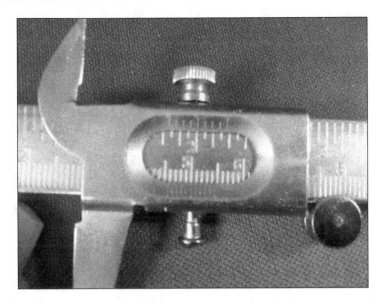

The leftmost tick is between 4.5 and 4.6, so this object is between 4.5 cm and 4.6 cm (or 45 mm/46 mm). Then count the tick marks on the outside until one of them lines up perfectly with one of the ticks on the inside. In this case, the fifth tick mark on the outside lines up with a tick mark on the inside. This is the hundredth part of the measurement so this object measures 4.55 cm or 45.5 mm.

Vernier calipers never need to be calibrated so they're never off because they weren't properly zeroed, they don't need batteries so they'll always work and they're cheaper. But they do take some additional effort to read properly and lack the "cool" factor of a seven segment display.

Keep in mind that if printing in ABS plastic, shrinkage after printing as discovered in *Chapter 4, SD Card Holder Ring*, will have to be taken into account. Also, no matter how accurately the measurements are taken, if the blueprints being developed involve strange angles or shapes there is a chance for inaccuracy translating that to the digital space.

Grid paper method

Fortunately, there is a way that a complex shape can be measured accurately no matter how strange the curves or turns. All that is needed is an object to measure, an ink pad, and some centimeter graph paper.

First thing to do is get an object that needs to be modeled for printing. In this example, a drawer guide that has an arm broken off will be used. If it were fixed, this piece should be identical on both sides. This piece has lots of complex shapes and measuring it completely would be a challenge.

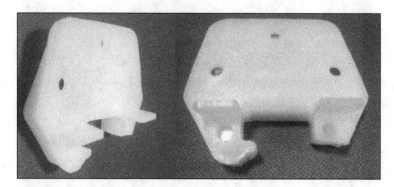

A good candidate for the grid paper method is a piece that has at least one flat side, or that can easily be made to have a flat side. The back of this piece is not flat at all, but the front has only two nubs sticking out that would be easy enough to remove. Just note their location before removing them since they'll need to be modeled back in. Then use a sharp blade and cut them off.

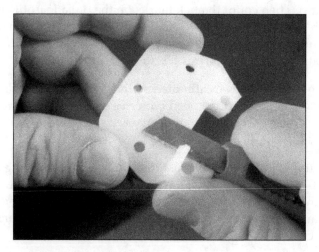

Then to be sure that the surface is perfectly flat, use some fine grit sandpaper on a flat surface to sand down the flat side.

Next, take some grid paper, centimeter if available, and an ink pad. Use the flat side of the object like a stamp. Try to line the part up with the grid as much as possible to make the later modeling process easier. Stamping on a soft surface like a towel can help get better coverage in the stamping process.

Finally, scan the stamped part in or if a scanner isn't available use a digital camera. Be sure that the grid is as lined up as possible, or some editing of the image will be necessary. It is not possible to edit the grid in Blender, so keep it straight horizontally and vertically, and avoid skewing and warping of the image as much as possible.

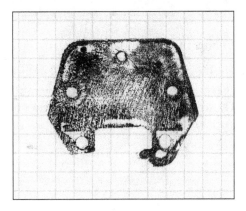

Using the grid paper method with Blender

To follow along from this point, download the previous screenshot from `http://thingiverse.com/thing:90754`. Perform the following steps to use the grid paper method in Blender:

1. Start Blender and as usual clear (*Ctrl + A, X*) the scene and save it to start a project. Give it an appropriate name in an appropriate directory such as `Ch10 Measuring and Drawer Guide.blend`.

2. Then change the view to **Top** (*Numpad 7*), **Orthographic** (*Numpad 5*).

3. Like in *Chapter 3, Face Illusion Vase*, in the Properties panel (*N*) locate the **Background Images** section, click on the checkbox next to it, expand it, and click on the **Add Image** button.

4. Click on the **Open Image** button and navigate to where the scanned image of the stamped object is stored. Then open the image to place in the scene.

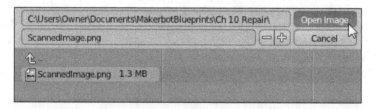

(The location and name of the scanned image may be different for the reader.)

The first thing to note is that if the previous steps were followed carefully the grid paper in the image is a centimeter grid, but that at the default zoom the grid lines seen are in millimeters. They are 10 millimeters per centimeter so zoom out until the major grid lines start to appear, then zoom out some more so that the millimeter grid lines disappear.

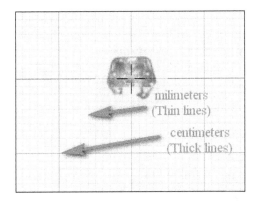

Adjust the settings in the Properties panel for the background image until the grid on the paper lines up with the centimeter grid in the 3D view. A quick way to determine the necessary size is to count the number of grid lines in the scan, multiply by 10, and divide by 2 (for the radius, which for some reason the size setting is). In the previous example, there are 12 grid lines shown so setting the size to 60 gets the view very close to the right size.

Continue to fiddle with the **X**, **Y**, and **Size** values until the grid lines in the image line up with the grid lines in the view properly.

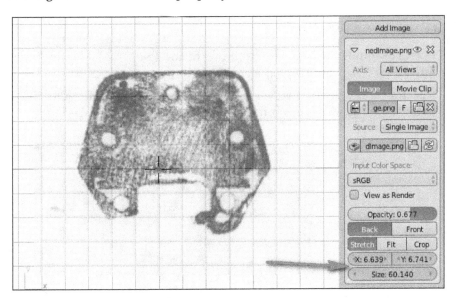

Since in this example the object is to be symmetrical it would probably be a good idea to use the **Mirror** modifier, and to use that properly it would be easier if the center line of the object was lined up with the world origin, so the **X** value will be adjusted to make that so. The **X** grid lines may no longer line up with the image, but that's less of a concern.

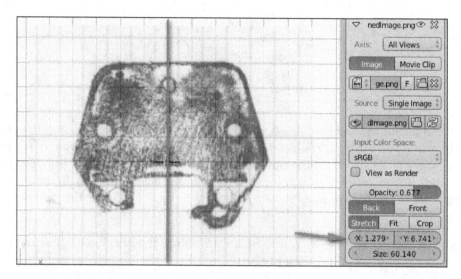

Now let's model the shape:

1. In this example, the best way to start is to add (*Shift + A*) a **Plane**, enter **Edit Mode** (*Tab*), and begin mesh editing.

2. Selecting and deleting two vertices, then extruding (*E*) one end of the line out to trace the shape, again similar to the way the face was traced for the face illusion vase.

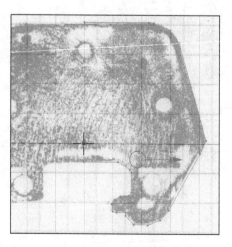

3. Add a **Mirror** modifier and click on the use modifier while in **Edit Mode** (the one that looks like a cube in the edit mode) just like in *Chapter 7, Teddy Bear Figurine.*

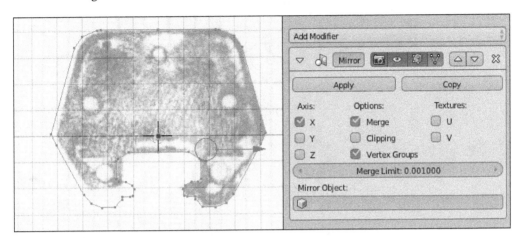

Once the **Mirror** modifier is added it is clear that the mesh is slightly off. The original image wasn't centered as well as it could have been. One way to fix this is to simply move all the points, until the model sticks out on one side as much as it overlaps on the other. The shape will be correct, but in the wrong place slightly relative to the source drawing.

Once the points are made they will need to be joined into a face and extruded to a shape:

1. Then exit **Edit Mode** (*Tab*), apply the **Mirror** modifier, and go back into edit mode (*Tab*).

2. Then select all the points (*A*) and create a new face (*F*).

3. If the shape isn't too complex and it seems to have worked, then nothing else is necessary to get a good starting shape. If not, then the `Create Face` function may need some help.

In this example, there are at least two problem areas. One area is where the face didn't fill and the other area where the face seems to have filled twice.

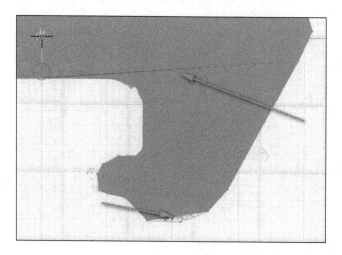

To fix this, whenever it comes up, make faces of smaller groups of points at a time.

1. **Circle Select** (*C*) smaller groups of adjacent vertices and create faces (*F*) for them.

2. Then deselect (*A*) the points and **Circle Select** (*C*) more points, including the last two of the previous face, and make a face (*F*) of them until full coverage is achieved. If a couple of points are causing trouble, a common problem with the **Mirror** modifier is duplicating points around the middle.

3. **Circle Select** (*C*) and **Merge** (*Alt + M*) them to clean up the mesh like in *Chapter 8, Repairing Bad Models*.

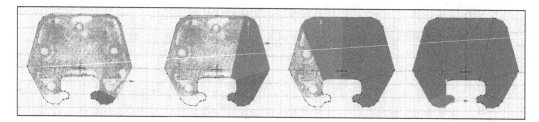

4. Once a solid shape is achieved select all (*A*) the vertices and **Extrude** (*E*) 9 mm to create a 3D shape.

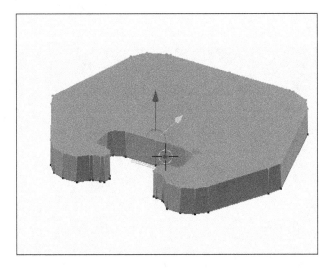

To finish the example in this chapter, the nubs that were cut off had to be added to the virtual model, so the cubes were added, then the **Boolean** modifier was be used to union them into the shape. The top of the cubes lined up with the bottom of the guide hole, and stood out about 8 mm from the flat surface. The holes were not reproduced in the copy since their presence was likely to reduce plastic usage in the injection molding process, not as much a problem for 3D printing with infill. Originally, the part was held on with wood staples but if necessary mounting holes can be drilled into the printed part to use screws.

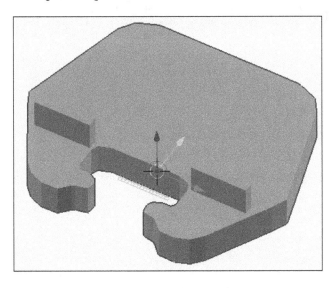

Since the scale of the object was determined by lining up the real life grid, the printed part should be approximately the same as the original piece, so all that is left is to navigate to **File | Export | Stl (.stl)**, export the object and print it out and compare it with the original.

The original object is still a little messy after being used as a stamp, but the new one will take its place nicely.

Full disclosure, this part is actually too big based on the provided grid. Centimeter graph paper was not available, so quarter inch graph paper was used instead. In order to make it work after modeling, the part simply needs to be scaled (*S*) down in the x and y axis (*Shift + Z*) by 0.63 units to fit the physical part. The shape is still perfect and the technique is sound, just the scale is wrong but fixable.

Summary

This final blueprint explored techniques for modeling 3D objects to fit when 3D printed. Measuring tools such as calipers are valuable aids to any 3D modeler, and using a scanner and ink pad is a clever technique to capture complex objects' shape, provided they have a flat side that can be used like a stamp. The grid paper method is a useful method that works in a surprising number of cases.

Perhaps, one day 3D scanners will be as ubiquitous as their 2D counterparts, but until that day finding ways to transfer real space into the digital space is a problem that will challenge 3D designers. Being creative about making accurate measurement will continue to be a rewarding challenge.

With this last technique, there should be nothing we need for designing objects for 3D printing. Congratulate yourself, you are now a 3D designer. Now the challenges are yours. Find objects that you want to make and model them, if you haven't already started. Find a need and fill it with a plastic object. Join an online community of designers such as Thingiverse and share your designs with others. Learn from their projects and models, continue to grow, and happy modeling!

Index

Thank you for buying
3D Printing Blueprints

About Packt Publishing

Packt, pronounced 'packed', published its first book *"Mastering phpMyAdmin for Effective MySQL Management"* in April 2004 and subsequently continued to specialize in publishing highly focused books on specific technologies and solutions.

Our books and publications share the experiences of your fellow IT professionals in adapting and customizing today's systems, applications, and frameworks. Our solution based books give you the knowledge and power to customize the software and technologies you're using to get the job done. Packt books are more specific and less general than the IT books you have seen in the past. Our unique business model allows us to bring you more focused information, giving you more of what you need to know, and less of what you don't.

Packt is a modern, yet unique publishing company, which focuses on producing quality, cutting-edge books for communities of developers, administrators, and newbies alike. For more information, please visit our website: www.packtpub.com.

Writing for Packt

We welcome all inquiries from people who are interested in authoring. Book proposals should be sent to author@packtpub.com. If your book idea is still at an early stage and you would like to discuss it first before writing a formal book proposal, contact us; one of our commissioning editors will get in touch with you.

We're not just looking for published authors; if you have strong technical skills but no writing experience, our experienced editors can help you develop a writing career, or simply get some additional reward for your expertise.

PUBLISHING

Blender 2.6 Cycles: Materials
and Textures Cookbook

Over 40 recipes to help you create stunning materials and
textures using the Cycles rendering engine with Blender

Enrico Valenza

Blender 2.6 Cycles: Materials and Textures Cookbook

ISBN: 978-1-78216-130-1   Paperback: 280 pages

Over 40 recipes to help you create stunning materials
and textures using the Cycles rendering engine
with Blender

1. Create naturalistic materials and textures - such
   as rock, snow, ice and fire - using Cycles

2. Learn Cycle's node-based material systems

3. Get to grips with the powerful Cycles
   rendering engine

Blender 3D Basics

The complete novice's guide to 3D modeling and animation

Beginner's Guide

Gordon C. Fisher

Blender 3D Basics

ISBN: 978-1-84951-690-7   Paperback: 468 pages

The complete novice's guide to 3D modeling
and animation

1. The best starter guide for complete newcomers
   to 3D modeling and animation

2. Easier learning curve than any other book
   on Blender

3. You will learn all the important foundation
   skills ready to apply to any 3D software

Please check **www.PacktPub.com** for information on our titles

Blender Game Engine: Beginner's Guide

ISBN: 978-1-84951-702-7   Paperback: 206 pages

The non programmer's guide to creating 3D video games

1. Use Blender to create a complete 3D video game

2. Ideal entry level to game development without the need for coding

3. No programming or scripting required

jMonkeyEngine 3.0 Beginner's Guide

ISBN: 978-1-84951-646-4   Paperback: 352 pages

Develop professional 3D games for desktop, web, and mobile, all in the familiar Java programming language

1. Create 3D games that run on Android devices, Windows, Mac OS, Linux desktop PCs and in web browsers – for commercial, hobbyists, or educational purposes

2. Follow end-to-end examples that teach essential concepts and processes of game development, from the basic layout of a scene to interactive game characters

3. Make your artwork come alive and publish your game to multiple platforms, all from one unified development environment

Please check **www.PacktPub.com** for information on our titles